T0224502

Maxwell's Equations

Jürgen Donnevert

Maxwell's Equations

From Current Density Distribution to the
Radiation Field of the Hertzian Dipole

 Springer Vieweg

Jürgen Donnevert
Dieburg, Germany

ISBN 978-3-658-29375-8   ISBN 978-3-658-29376-5   (eBook)
https://doi.org/10.1007/978-3-658-29376-5

Responsible Editor: Reinhard Dapper
This Springer Vieweg imprint is published by the registered company Springer Fachmedien Wiesbaden GmbH part of Springer Nature.
The registered company address is: Abraham-Lincoln-Str. 46, 65189 Wiesbaden, Germany

Preface

James Maxwell[1] postulated: "A time varying electric field is linked to a time varying magnetic field and vice versa, even without the presence of a conductor." This insight is the basis for Maxwell's equations that are differential equations describing the nature of electric and magnetic fields and the existence of electromagnetic waves.

This volume focuses on the derivation and solution of Maxwell's equations. The book was chiefly written for students of electrical engineering, information technology, and physics, the goal being to prepare them for courses on Electromagnetic Field Theory (EFT). Building of what they have learned in advanced physics and mathematics courses at grammar schools and comprehensive schools, it is intended to accompany EFT courses. Particular importance is attached to detailed explanations in text form, combined with a wealth of illustrations. All formulas are derived step by step.

The stations on the way to Maxwell's equations include the laws of the current density distribution, as well as the laws of electrostatics and magnetostatics. The first steps are basic experimental arrangements of these sub-areas of the field theory. In the first chapters, the terms scalar field and vector field are introduced. The vector-analytical operators gradient, divergence, and curl, which are required for the description and calculation of these fields, are derived for three coordinate systems and explained with examples, as well as the integral theorems of Gauss and Stokes. Chap. 3 deals with the stationary magnetic field, and the magnetic vector potential is introduced.

Subject of Chap. 4 are time-varying fields. In this chapter, Faraday's law of induction is deduced which results in the second Maxwell's equation. In the next step, the continuity equation is formulated and the displacement current is introduced. The addition of the displacement current to Faraday's law leads to Maxwell's first equation. The two Maxwell's equations in integral and differential form are discussed. Hereafter, the special form of Maxwell's equations for time-harmonic dependence is presented. The general form of the homogeneous and inhomogeneous wave equation shows that the wave

---

[1]Maxwell, James Clerk, British Physicist, *1836, †1879.

equation also applies to the electric and magnetic field vectors and the magnetic vector potential. The solution of Maxwell's equations by the retarded vector potential is given. The chapter ends with Poynting's vector, describing the energy flux of electromagnetic fields.

The focus of Chap. 5 is the Hertzian dipole. This example shows in detail the propagation of electromagnetic waves. The field equations for the near and far fields are derived, and field lines of the electromagnetic far field are given. Finally, important characteristic parameters of antennas such as radiation pattern, antenna gain, and effective area are discussed. Important equations and text are highlighted by shading the text.

This volume introduces the reader step by step to the fascinating world of electromagnetic field theory.

Dieburg   Jürgen Donnevert
December 2019

Acknowledgements My thanks go to Mr. Joachim Elser for his references to misprints and to Mr. Michael Lenz for his comments on the induction law. The entire manuscript was reviewed by Professor Dr. Heinz Schmiedel and Professor Dr. Manfred Götze, who made significant contributions to the content as well as to the style of this work.

Contents

List of abbreviations

$\vec{J}_D$	Displacement current density, vector
P_1, P_2	Power 1, Power 2
P_{rec}	Receive power
P_{tr}	Transmit power
P_{rad}	Radiated power
P_{act}	Active power
R_{rad}	Real part of the input impedance of an antenna, radiation resistance
R_H	Hall constant
R_L	Real part of the impedance of the receiver
S_{rec}	Power density at the receiving location
$S_{Hertz/max}$	Maximum power density in the far field of the Hertzian dipole
$S_{isotrop}$	Power density of the isotropic radiator
V_{12}	Voltage between electrodes 1 and 2
V_H	Hall voltage
V_{ab}	Voltage between locations a and b
X_A	Imaginary part of the internal antenna impedance
X_L	Imaginary part of the input receiver impedance
Z_0	Wave impedance
a_0	Free-space loss
a_{RF}	Radio field loss
c_0	Speed of light
g_{ant}	dB-gain of an antenna
g_{Hertz}	dB-gain of the Hertzian dipole
g_{tr}	dB-gain of the transmitting antenna
g_{rec}	dB-gain of the receiving antenna
i_D	Displacement current
k_0	Wave number
t^*	Retarded time
v_{12}	Voltage between terminals 1 and 2, e.g., of an inductor
w_{el}	Energy density of the electric field
w_{magn}	Energy density of the magnetic field
A	Surface, enveloping surface, ampere (unit)
$A = \left|\vec{A}\right|$	Magnetic vector potential
B	Magnetic flux density
C	Coulomb, capacitance of a capacitor, capacitor
C	Contour
D	Diameter
D	Electrical flux density, displacement density
E	Amplitude of the electric field
F	Force
H	Henry, unit

H	Magnetic field
I	Electric current, current intensity
K	Proportionality factor, constant
L	Inductance
N	Newton, unit
N	Number of turns
P	Power
Q	Electrical charge
Q	Deflection of the ballistic galvanometer
R	Resistor
T	Time period, Tesla (unit)
V	Voltage
V	Volt, unit, volume
W	Energy, watt, unit
Wb	Weber, unit
b	Width
c	Velocity of propagation
cm	Centimeter
d	Distance, distance between the transmitting antenna and the receiving antenna, radio field length
dA	Surface element, infinitesimal current
dA_r	Surface element, oriented perpendicular to the r-direction
dI	Infinitesimal current intensity
dP	Infinitesimal power
dQ	Infinitesimal charge, respectively, charge quantity
dV	Volume element, infinitesimal volume
dW	Energy in the volume element dV, energy absorbed or dissipated during time period dt
dn	Infinitesimal path difference, infinitesimal distance
ds	Path or length element, infinitesimal length, or distance
dt	Infinitesimal time period
dv	Infinitesimal change of the voltage
e	Charge of an electron
f	Frequency
f	Function designation
g	Function designation
h	Height
i	Instantaneous value of the electric current
j	Designation of the imaginary part of a complex number or of a complex vector with time-harmonic dependence

l	Length
$\log_{10}$	Logarithm to base 10
m	Meter, Unit
n	Number of windings per length unit
p	Power density
q	Instantaneous value of the charge
q	Efficiency of an aperture antenna
r	Radius
s	Second
t	Time
v	Instantaneous value of the voltage, designation of a function, speed
w	Function designation

Greek letters

Φ_{con}	Concatenated magnetic flux
ε_o	Electrical field constant, dielectric conductivity of vacuum, absolute permittivity
ε_r	Relative permittivity
μ_0	Absolute permeability, permeability of the vacuum
μ_r	Relative permeability
$\underline{\varphi}$	Scalar potential, phasor, time-harmonic dependence
φ_0	Phase angle
φ_a	Potential of the potential surface a
φ_b	Potential of the potential surface b
ΔA	Small surface
ΔQ	Small charge
$\nabla \varphi$	Nabla operator to be applied to the scalar potential field φ, gradient of the scalar potential field φ
Φ	Magnetic flux
$d\Phi_{con}$	Infinitesimal magnetic flux, concatenated with a conductor loop
$d\varphi$	Infinitesimal change of the angle φ
α	Angle
β	Angle
ε	Permittivity, dielectric conductivity
κ	Specific resistance
λ	Wavelength
μ	Permeability
π	Number Pi
σ	Specific conductivity
φ	Angle, scalar potential, scalar potential field

$\varphi(x, y, z)$	Three-dimensional scalar potential field
ψ	Scalar magnetic potential
ω	Angular frequency
ϑ	Angle
ϱ	Space charge density

Mathematical formula symbols and operators

$\int_a^b$	Integral along the path between the points a and b of a field or function
$\oint_C$	Enclosed integral via the contour or loop along C
$\oint$	Integral over a closed path
$\iint_A$	Integral over the surface A
$\oiint_A$	Enclosed integral over the surface A closed in itself
$\iiint_V$	Integral over the volume V
$\sum_{j=0}^{N}$	Sum from $j = 0$ to N
$\underline{\vec{E}}$	Vector of the electric field, phasor, time-harmonic dependence
$\nabla^2 \varphi$	$\nabla \cdot \nabla \varphi = \text{div}(\text{grad } \varphi)$
$\text{curl}_r \vec{H}$	r-component of $\nabla \times \vec{H}$, r-direction
$\vec{A} \cdot \vec{B}$	Scalar product of the vectors $\vec{A}$ and $\vec{B}$
$\frac{\partial}{\partial x}$	Partial derivative with respect to x
$\nabla \cdot \vec{D}$	div $\vec{D}$
$\nabla \times \vec{D}$	Curl of the vector $\vec{D}$
$\nabla \varphi$	Nabla operator applied to the scalar potential field φ, grad φ
div $\vec{D}$	Divergence of the electric flux density vector $\vec{D}$
grad φ	$\nabla \varphi$, Gradient of the scalar field φ
curl $\vec{H}$	$\nabla \times \vec{H}$, curl of the magnetic field vector $\vec{H}$
$\text{Re}\{\underline{\vec{E}}\}$	Real part of the electric field, phasor, time-harmonic dependence

Vectors

$\underline{\vec{A}}_z$	Z-component of the vector potential, phasor, time-harmonic dependence
$\underline{\vec{H}}^*$	Complex conjugate vector of the magnetic field, phasor, time-harmonic dependence

$\vec{\underline{A}}$	Vector of the magnetic vector potential, time-harmonic dependence
$\vec{\underline{E}}$	Vector of the electric field, phasor, time-harmonic dependence
$\left\|\vec{\underline{E}}\right\|$	Magnitude of the vector $\vec{E}$
$\vec{\underline{J}}$	Vector of the current density, phasor, time-harmonic dependence
$\vec{\underline{J}}_n$	Component of the vector of the current density, perpendicular to the surface element $\mathrm{d}A$
$\vec{\underline{J}}_t$	Component of the vector of the current density, tangential to the surface element $\mathrm{d}A$
$\vec{\underline{J}}_D$	Displacement current density
$\vec{\underline{S}}$	Vector of the power flux density, Poynting vector
$\vec{\underline{S}}_{\text{active}}$	Vector, active power density in the direction of the Poynting vector
$\vec{e}_r$	Unit vector in r-direction, radial direction
$\vec{e}_x$	Unit vector, x-direction
$\vec{e}_y$	Unit vector, y-direction
$\vec{e}_z$	Unit vector, z-direction
$\vec{e}_\alpha$	Unit angle vector in α-direction
$\vec{e}_\vartheta$	Unit angle vector in ϑ-direction
$\vec{n}_{\Delta A}$	Unit vector, perpendicular to the infinitesimal surface ΔA
$\vec{n}_A$	Unit vector, perpendicular to the surface A
$\vec{\underline{s}}$	Vector, vector field, phasor, time-harmonic dependence, interference vector
$\vec{\underline{w}}$	Vector function, vector field, phasor, time-harmonic dependence
$\vec{A}$	Vector potential
$\vec{B}$	Vector of the magnetic flux density
$\vec{D}$	Vector of the electrical flux density or of the displacement density
$\vec{E}$	Vector, electric field
$\vec{F}$	Force vector
$\vec{H}$	Vector of the magnetic field
$\vec{J}$	Vector of the current density
$\vec{S}$	Poynting vector
$\vec{V}$	Vector, general
$\vec{l}$	Vector of a length
$\vec{r}$	Radius vector, distance vector to the origin of the coordinate system.
$\vec{v}$	Velocity vector
$\mathrm{d}\vec{A}$	Vector of the surface element (infinitesimal) which is perpendicular to the surface element.
$\mathrm{d}\vec{s}$	Vector of the path element ds

Potential and Current Density Distribution

1

We will begin with the description of a simple experiment where we will study the flow of steady electric current or more precisely, the current density. The measuring arrangement is shown in Fig. 1.1. It consists of a glass container with two electrodes connected to a DC voltage source. The container is filled with tap water. The water is conductive. The test is carried out with a voltage of 24 V[1]. The arrangement is homogeneous in z-direction.

For the test, electrode *A* is connected to the negative electrode of a high-impedance voltmeter, while the other electrode (measurement sensor) is moving in the water and the voltage is measured at as many points as possible. Figure 1.1 shows the measurement results as curves with the same measured values as solid lines. These lines are lines of equal potential. In this experiment, they form the edge of surfaces of equal potential extending from the water surface to the bottom of the glass container. Since the two electrodes also extend from the water surface to the bottom of the glass container, the value of the potential in this particular case does not depend on the z-coordinate.

---

[1]V = Volt is the unit of measurement defined in the International System of Units (SI) for electrical voltage. It was named after the Italian physicist Alessandro Volta. The capital letter "V" is used as the unit character.

The volt is a derived SI unit. With the SI base units Watt (W) and Ampere (A), we get

$$1\,V = 1\,\frac{W}{A} = 1\,\frac{N\,m}{A\,s} = 1\,\frac{kg\,m^2}{A\,s^3} = 1\,\frac{kg\,\frac{m^2}{s^2}}{A\,s} = 1\,\frac{kg\cdot\frac{m^2}{s^2}}{A\cdot s}$$

As this definition can hardly be used for calibration purposes as an exact reference, since 1990 the unit Volt is determined by the Josephson effect and the Josephson constant. The unit Ampere (A) is introduced in Sect. 3.2.1.

Historically, the definition of a Volt was derived from the Weston normal element. This element supplies an electrical voltage of exactly 1.01865 V at a temperature of 20°C.

© Springer Fachmedien Wiesbaden GmbH, part of Springer Nature 2020
J. Donnevert, *Maxwell´s Equations*, https://doi.org/10.1007/978-3-658-29376-5_1

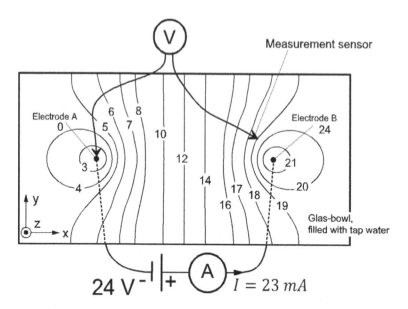

Fig. 1.1 Experiment: Measurement of equipotential surfaces (horizontal section through the surfaces with equal potential). Numbers show the voltage in V

The reference point for the potentials in Fig. 1.1 is electrode A. This reference point for the potential is arbitrary. If the reference point is changed, the shapes of the potential surfaces will not change, only the measured values of the voltmeter, i.e., the values of the equipotential surfaces will change. If, for example, the reference point is set to the 6 V equipotential line, all potentials are reduced by 6 V (see Fig. 1.2). If both sensors are set to the same potential, the display will show 0 V.

The contours of the surfaces with equal potential shown in Figs. 1.1 and 1.2 are part of a scalar potential field[2]. The potential of a point in the potential field is equal to the voltage between this point and a reference point. Consequently, the voltage between any two points of the potential field is equal to the potential difference between these points.

Designating the potential of point a with φ_a and the potential of point b with φ_b, then the voltage V_{ab} between these points is

$$V_{ab} = \varphi_a - \varphi_b \qquad (1.1)$$

A current flows between the two electrodes A and B, which happens to be 23 mA during a test[3]. The current flows over the entire space filled with water in the glass container.

---

[2]A scalar field is a function that assigns a real number (scalar) to each point in a space.

[3]Ampere (A) unit is one of the four basic units of the SI international system of units. The definition of this unit is discussed in Sect. 3.2.1.

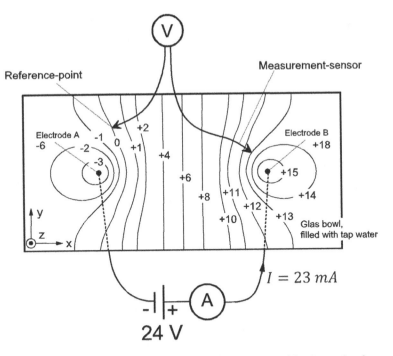

Fig. 1.2 Horizontal section through the equipotential surfaces with changed reference point. Numbers show the voltage in V

The positive direction of the current, which is indicated in Figs. 1.1 and 1.2 by the arrow, is determined by the following agreement:

The current flows from locations with higher potential to locations with lower potential.

This definition is based on the assumption that the electric current represents a movement of positively charged carriers. However, since the electric current is produced by a movement of electrons moving due to their negative charge from locations of lower potential to locations of higher potential, this convention defines the positive current direction opposite to the direction of the real movement of the electrons.

Characteristics for the current flow in space is the current density at the point in space under consideration. The current density is the quantity of charge moving per unit time through the surface in question, i.e., the current per unit area. If the area is close to zero, we will obtain the current density for the point of the surface in question. In contrast to the potential, the current density is a vector pointing into the direction of the flow of current. Its magnitude corresponds to the value of the current density. As a result, the field of the current density is a vector field[4]. In the following text, the current density is

---

[4]A vector field is a function that assigns a vector to each point in a space.

Fig. 1.3 Current density

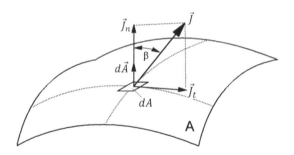

designated as $\vec{J}$. The unit of current density is A/m^2. The arrow above the letter J indicates that it is a vector.

To explain the relationship between current density and current, a curved surface through which a current flows is shown in Fig. 1.3. The vector $\vec{J}$ is the vector of the current density of the surface element dA. Since the surface element is assumed to be very small, the current density in the surface element dA is constant. The vector d$\vec{A}$ is a vector which is perpendicular to the surface element dA. Its magnitude corresponds to the surface of the element dA. The current density vector $\vec{J}$ is divided into a component $\vec{J}_t$ which is tangential to the surface, and a component $\vec{J}_n$ which is, like the vector d$\vec{A}$, perpendicular to the surface element dA. Only the part $\vec{J}_n$ of the current density penetrates the surface A.

The current dI passing through the surface element dA is

$$dI = \left|\vec{J}_n\right| \cdot \left|d\vec{A}\right|$$

$$dI = \left|\vec{J}\right| \cdot \cos\beta \cdot \left|d\vec{A}\right|$$

i.e.,

$$dI = \vec{J} \cdot d\vec{A} \tag{1.2}$$

The multiplication point in (1.2) designates the scalar product of the two vectors $\vec{J}$ and d$\vec{A}$.

The total current passing through the surface A is obtained by integration over this surface A

$$I = \iint_A \vec{J} \cdot d\vec{A} \tag{1.3}$$

The double integral indicates an integration over a surface.

Explanation: Scalar Product

The scalar product is a product of two vectors. The result is a scalar[5]. For the calculation of the scalar product, the vectors must have the same number of components.

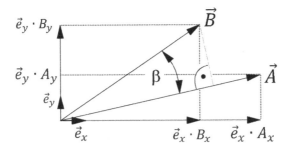

The figure shows two vectors $\vec{A}$ and $\vec{B}$ which enclose the angle β in the x-y plane with their components A_x, A_y, B_x, and B_y. In addition, the unit vectors $\vec{e}_x$ and $\vec{e}_y$ are depicted. With

$$|\vec{e}_x| = \vec{e}_x \cdot \vec{e}_x = 1$$

$$|\vec{e}_y| = \vec{e}_y \cdot \vec{e}_y = 1$$

$$\vec{e}_x \cdot \vec{e}_y = 0$$

we obtain the scalar product of the two vectors $\vec{A}$ and $\vec{B}$ as

$$\vec{A} \cdot \vec{B} = \left(A_x \cdot \vec{e}_x + A_y \cdot \vec{e}_y\right) \cdot \left(B_x \cdot \vec{e}_x + B_y \cdot \vec{e}_y\right)$$

$$\vec{A} \cdot \vec{B} = (A_x \cdot \vec{e}_x) \cdot (B_x \cdot \vec{e}_x) + (A_x \cdot \vec{e}_x) \cdot \left(B_y \cdot \vec{e}_y\right) + \left(A_y \cdot \vec{e}_y\right) \cdot (B_x \cdot \vec{e}_x) + \left(A_y \cdot \vec{e}_y\right) \cdot \left(B_y \cdot \vec{e}_y\right)$$

$$\vec{A} \cdot \vec{B} = (A_x \cdot B_x \cdot \vec{e}_x \cdot \vec{e}_x) + \left(A_x \cdot B_y \cdot \vec{e}_x \cdot \vec{e}_y\right) + \left(A_y \cdot B_x \cdot \vec{e}_y \cdot \vec{e}_x\right) + \left(A_y \cdot B_y \cdot \vec{e}_y \cdot \vec{e}_y\right)$$

$$\vec{A} \cdot \vec{B} = A_x \cdot B_x + A_y \cdot B_y$$

The following equation leads to the same result:

$$\vec{A} \cdot \vec{B} = \left|\vec{A}\right| \cdot \left|\vec{B}\right| \cdot \cos\beta$$

---

[5]A scalar is a mathematical quantity that is characterized solely by a numerical value.

For vectors with x-, y-, and z-components

$$\vec{A} \cdot \vec{B} = A_x \cdot B_x + A_y \cdot B_y + A_z \cdot B_z$$

In text form:

The scalar product of two vectors is the sum of the products of the corresponding components.

In Fig. 1.4, the flow of current in the glass bowl between the electrodes A and B is represented by current density lines. In addition to the current density lines, the cross sections of the equipotential surfaces, i.e., the lines of equal potential, are also drawn in this figure. In the experiment, the current flows from electrode B distributed over the entire space in the glass container to electrode A. Since there is no potential difference along the equipotential surfaces, the direction of the current density vector in each point of the potential field is orthogonal, i.e., at right angle, to the potential surfaces.

1.1   Electric Field

The cause of the current flow in the test arrangement, according to Fig. 1.4, is the voltage between the two electrodes. An increase in voltage between the electrodes results in an increase of the current. The potential in the special case of Fig. 1.4 does not depend on the z-coordinate. Therefore, a three-dimensional problem can be reduced to a two-dimensional problem. In Fig. 1.4, we consider a small volume. The side surfaces of this volume are

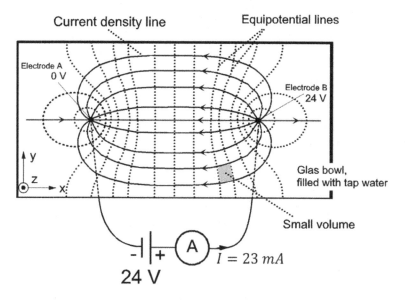

Fig. 1.4 Horizontal section through the equipotential surfaces (*dotted*) and current density lines (*solid lines*)

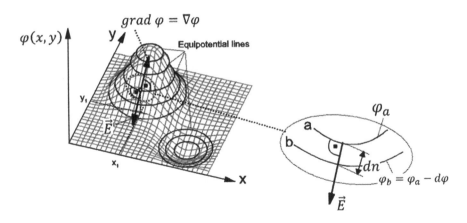

Fig. 1.5 Two-dimensional potential field

surfaces of equal potential. The current density within the volume depends on the potential difference in the side surfaces. As a next step, this volume is transformed into an infinitesimal volume. To describe mathematical the current density in the point of the infinitesimal volume, a new quantity has to be defined: the electric field. The electric field is a vector which, like the current density, is perpendicular to the equipotential surfaces and points into the direction of the lower potential. The magnitude of the electric field is obtained from the difference $d\varphi$ of the potential along the infinitesimal distance dn of two infinitesimal equipotential surface elements divided by their distance dn. Like the current density vectors, the vectors of the electric field also form a vector field in space.

For further explanation of the term "electric field", a section of a two-dimensional potential field is shown in Fig. 1.5. The potential values $\varphi(x, y)$ are shown in the form of a potential mountain. The mesh structure in the figure is required to create a three-dimensional impression.

In the right part of Fig. 1.5, a section of the potential field with two equipotential lines is shown. The potential difference $\varphi_a - \varphi_b$ of two equipotential lines with infinitesimal distance dn is designated as $d\varphi$. The magnitude of the electric field is

$$\left|\vec{E}\right| = E = \frac{\varphi_a - \varphi_b}{dn} = \frac{d\varphi}{dn} \tag{1.4}$$

By definition, $d\varphi$ is positive for $\varphi_a > \varphi_b$. The electric field has the dimension V/m.

To illustrate the electric field as a vector field, the vectors of this field are shown in Fig. 1.6. At each point of a field line, the tangent to the field line matches the direction of the vector in that field point. Since the distance between the equipotential lines decreases, the closer they are to the two electrodes, the electric field increases, the closer the observed point is to one of the two electrodes.

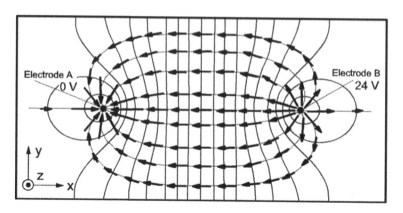

Fig. 1.6 Scalar potential and vector field of the electric field

In vector analysis, the vector pointing into the direction of the largest increase of a scalar field is called a gradient.[6] Since the vector of the electric field is by definition in the direction of the strongest decrease of the scalar potential field $\varphi(x, y, z)$, we have

$$\vec{E} = -\text{grad}\,\varphi \tag{1.5}$$

Instead of the notation *grad* φ, the notation $\nabla\varphi$ is common. The symbol ∇ is referred to as the Nabla operator (pronounced "del")

$$\vec{E} = -\text{grad}\,\varphi = -\nabla\varphi \tag{1.6}$$

1.1.1 Explanation: Gradient or Nabla Operator

The gradient (grad or ∇) is a mathematical differential operator applied to a scalar local function, in the resent case to a potential field. The result is a vector field (cf. (1.6)). In the Cartesian coordinate system, this means

$$\nabla = \frac{\partial}{\partial x} \cdot \vec{e}_x + \frac{\partial}{\partial y} \cdot \vec{e}_y + \frac{\partial}{\partial z} \cdot \vec{e}_z$$

respectively

$$\nabla\varphi = \text{grad}\,\varphi = \frac{\partial\varphi}{\partial x} \cdot \vec{e}_x + \frac{\partial\varphi}{\partial y} \cdot \vec{e}_y + \frac{\partial\varphi}{\partial z} \cdot \vec{e}_z \tag{1.7}$$

---

[6]Lat. gradus = step.

Fig. 1.7 Unit vectors in the cylindrical-coordinate system

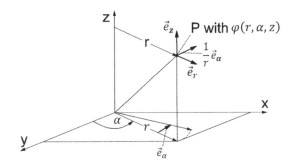

The gradient of a potential field is thus determined as follows:

1. Determine the variation of the potential $\varphi(x, y, z)$ when progressing in x-direction by an infinitesimal step and multiplication with the unit vector $\vec{e}_x$ which points in x-direction.
2. Perform the described operations analogously for the y- and z-directions.
3. Add the three vectors.

The procedure described above defines a vector pointing into the direction of the largest increase in the potential field.

Cylinder Coordinates

For cylinder-symmetric problems, it is useful to use the vector operation $\nabla \varphi$ in the cylinder-symmetric coordinate system.

With regard to the r- and z-coordinates, we proceed as in the case of the Cartesian coordinates (see Fig. 1.7). When determining the α-coordinate, note that the magnitude of the unit angle vector $\vec{e}_\alpha$ points into the direction of the tangent of the circle with the radius r. A change of an angle in the x-y plane results in a change of length on the arc of the circle which depends on the radius r of the circle. A change of the angle α of 1 rad[7] results in a change in length $1 \cdot r$ LU[8] on the arc of the circle with the radius r. The unit length vector $\vec{e}_{arc}$ pointing into the direction of the tangent of the circle radius r is equal to 1 independent of the position of the point P

$$\vec{e}_{arc} = \left| \frac{1}{r} \cdot \vec{e}_\alpha \right| = 1$$

---

[7]$360° = 2 \cdot \pi \; rad.$
[8]$LU =$ Length unit.

Fig. 1.8 Unit vectors in the
spherical coordinate system

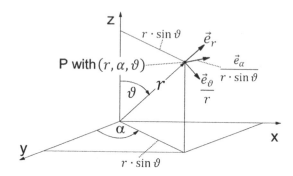

Execution of the operation $\nabla\varphi(r,\alpha,z) = \mathrm{grad}\,\varphi(r,\alpha,z)$ in four steps:

1. Variation of potential $\varphi(r,\alpha,z)$ in point P while progressing by an infinitesimal step into r-direction and multiplication with the unit vector $\vec{e}_r$ pointing into r-direction.
2. Variation of potential $\varphi(r,\alpha,z)$ while progression by an infinitesimal step into α-direction and multiplication with the unit length vector defined in α-direction, i.e., with the vector $\vec{e}_\alpha/r$.
3. Variation of potential $\varphi(r,\alpha,z)$ in point P while progression by an infinitesimal step in z-direction and multiplication with the unit vector $\vec{e}_z$ pointing into z-direction.
4. Addition of the three parts.

Thus the gradient in cylindrical coordinates operation is[9]

$$\nabla\varphi = \mathrm{grad}\,\varphi = \frac{\partial\varphi}{\partial r}\cdot\vec{e}_r + \frac{1}{r}\cdot\frac{\partial\varphi}{\partial\alpha}\cdot\vec{e}_\alpha + \frac{\partial\varphi}{\partial z}\cdot\vec{e}_z \qquad (1.8)$$

Spherical Coordinates
For spherical coordinates, the same applies as for the unit vector $\vec{e}_\alpha$ for cylinder coordinates (see Fig. 1.8).

Unit length vector in α-direction: $\frac{1}{r\cdot\sin\vartheta}\cdot\vec{e}_\alpha$
Unit length vector in ϑ-direction: $\frac{1}{r}\cdot\vec{e}_\vartheta$
Unit length vector in r-direction: $\vec{e}_r$

In spherical coordinates, $\mathrm{grad}\,\varphi = \nabla\varphi$ is[10]

$$\nabla\varphi = \mathrm{grad}\,\varphi = \frac{\partial\varphi}{\partial r}\cdot\vec{e}_r + \frac{1}{r\cdot\sin\vartheta}\cdot\frac{\partial\varphi}{\partial\alpha}\cdot\vec{e}_\alpha + \frac{1}{r}\cdot\frac{\partial\varphi}{\partial\vartheta}\cdot\vec{e}_\vartheta \qquad (1.9)$$

---

[9]For $r = 0$, the α-component is obviously not defined.
[10]For $r = 0$ and $\vartheta = 0$ the α- and ϑ-components are obviously not defined.

1.2   Kirchhoff's Laws

The first of the two laws for electrical networks formulated by Kirchhoff[11]:

> The algebraic sum of the incoming currents at a point (junction or node) of an electrical network is equal to the sum of the outgoing currents.

When applied to the field of the electric current densities, this law has to be modified. In the field of electric current densities, a node has to be replaced by a closed surface. In Fig. 1.9, such a surface A is depicted with five infinitesimal surface elements dA and five current density vectors $\vec{J}_1$ to $\vec{J}_5$. According to (1.2), a current I passing through a surface element is $\vec{J} \cdot d\vec{A}$. The vector $d\vec{A}$ is perpendicular to the surface element dA and directed outward. According to Fig. 1.3 and (1.2),

$$\vec{J}_1 \cdot d\vec{A} = I_1$$

and

$$\vec{J}_2 \cdot d\vec{A} = I_2$$

are negative currents. The currents enter the closed surface A. The currents

$$\vec{J}_3 \cdot d\vec{A} = I_3$$

$$\vec{J}_4 \cdot d\vec{A} = I_4$$

and

$$\vec{J}_5 \cdot d\vec{A} = I_4$$

Fig. 1.9 About the first Kirchhoff's law

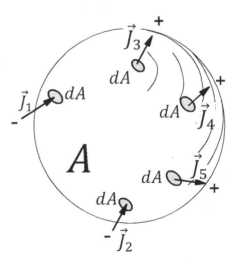

---

[11]Gustav Robert Kirchhoff, German physicist, *1824, †1887.

are positive currents because they are leaving the surface A. The paths of the currents within the closed surface are not drawn. There is no current source inside the closed surface. Therefore, the sum of the currents entering the surface is equal to the sum of the currents leaving the surface, i.e.,

$$\sum_{n=1}^{5} \vec{J}_n \cdot \mathrm{d}\vec{A} = 0 \tag{1.10}$$

In a field, the current densities are not restricted to narrow paths. They are generally distributed throughout the entire space. The sum in (1.10) has therefore to be replaced by an integral.

$$\oiint_A \vec{J} \cdot \mathrm{d}\vec{A} = 0 \tag{1.11}$$

The ellipse around the double integral sign in (1.11) indicates that the surface A is closed. Equation (1.11) is the first Kirchhoff´s law for fields of current density.

The second Kirchhoff´s law for networks:

> All partial voltages of a circuit or a mesh in an electrical network add up to zero. The direction of rotation can be selected at will. Partial voltages in the direction of rotation are positive, partial voltages against the direction of rotation are negative.

In a potential field, instead of the voltage, the electrical field is integrated along a closed path. Figure 1.10 shows such a path from point a to point b and back to point a. The lower part of the figure shows an enlarged section with this closed path.

After (1.4), the electric field is equal to the potential difference $\boldsymbol{d\varphi}$ divided by the distance between the equipotential surfaces

$$\left|\vec{E}\right| = \frac{\mathrm{d}\varphi}{|\mathrm{d}\vec{s}| \cdot \cos\beta}$$

respectively

$$\mathrm{d}\varphi = \left|\vec{E}\right| \cdot |\mathrm{d}\vec{s}| \cdot \cos\beta$$

If $\mathrm{d}\varphi$ is positive, the potential will increase by advancing into the direction of $\mathrm{d}\vec{s}$. Since the vector of the electric field by definition points into the direction of the decreasing potential, we get

$$\mathrm{d}\varphi = -\vec{E} \cdot \mathrm{d}\vec{s} \tag{1.12}$$

Thus the voltage V_{ab} between a and b in Fig. 1.10, by advancing into the direction of $\mathrm{d}\vec{s}$ (starting point $= a$), is

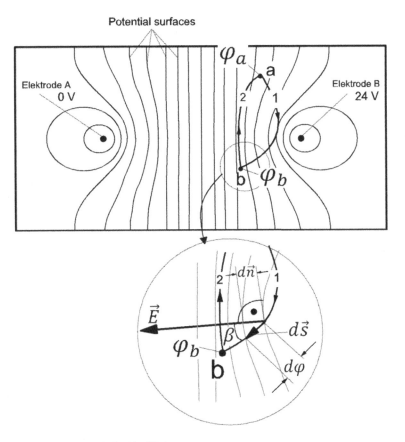

Fig. 1.10 About the second Kirchhoff's law

$$V_{ab} = -\int_b^a \vec{E} \cdot d\vec{s} = -(\varphi_a - \varphi_b) = \varphi_b - \varphi_a \qquad (1.13)$$

The integral starting from point b toward the point a has the same magnitude but an opposite sign:

$$V_{ba} = -\int_a^b \vec{E} \cdot d\vec{s} = -(\varphi_b - \varphi_a) = \varphi_a - \varphi_b \qquad (1.14)$$

The sum of the two integrals in (1.13) and (1.14) is therefore equal to zero. This means that the integral of the electric field on a closed path will be zero, if the path contains no voltage sources

$$\oint \vec{E} \cdot d\vec{s} = 0 \qquad (1.15)$$

This is the formulation of the second Kirchhoff's law for the electric field.

Putting (1.6)

$$\vec{E} = -\text{grad } \varphi = -\nabla\varphi$$

into (1.13)

$$V_{ab} = -\int_b^a \vec{E} \cdot d\vec{s} = \int_a^b \text{grad } \varphi \cdot d\vec{s} = \varphi_b - \varphi_a \qquad (1.16)$$

Interpretation of Eq. (1.16) and (1.15):

→ The integral of the electric field between two points a and b is equal to the potential difference between these two points and does not depend on the path of integration.

1.3   Ohm´s Law in the Field of Current Densities

Figure 1.11 shows two potential surfaces and a small prism. The distance between the front surfaces of the prism is dn. The front surfaces of the prism are located on potential surfaces. The lateral surfaces of the prism are perpendicular to the potential surfaces, so that the current density lines are parallel to these surfaces. Current dI enters into the front surface dA. Since the surface dA is small, the current density J will be

$$J = \frac{dI}{dA} \qquad (1.17)$$

The Ohmic resistance of the prism is inversely proportional to the specific conductivity [12]σ and to the cross-sectional surface dA, and proportional to the length dn of the prism. Thus the resistance R of the prism is[13]

$$R = \frac{dn}{\sigma \cdot dA} \qquad (1.18)$$

With Ohm´s law, we obtain the voltage drop $d\varphi$ along the prism

$$d\varphi = R \cdot dI = \frac{dn}{\sigma \cdot dA} \cdot dI \qquad (1.19)$$

$$\frac{d\varphi}{dn} = \frac{dI}{\sigma \cdot dA} \qquad (1.20)$$

---

[12]Unit of the specific conductivity $[\sigma]$: $\frac{S \cdot m}{mm^2}$ but mostly $\frac{S}{m}$ (S = Siemens = $\frac{1}{\Omega}$).
[13]Ohm's law.

Fig. 1.11 About Ohm's law in the field of current densities

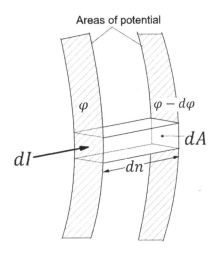

Areas of potential

With (1.4)

$$\left|\vec{E}\right| = \frac{d\varphi}{dn}$$

and (1.17) follows from (1.20)

$$\left|\vec{E}\right| = \frac{1}{\sigma} \cdot \left|\vec{J}\right| = \kappa \cdot \left|\vec{J}\right| \tag{1.21}$$

κ = specific electric resistance[14]

Since the vector of the electric field points into the direction of the vector of the current density[15], we get

$$\vec{E} = \frac{1}{\sigma} \cdot \vec{J} = \kappa \cdot \vec{J} \tag{1.22}$$

1.4   Power Density in the Field of Current Densities

In a conductor carrying the current I and to which the voltage V is applied, a power P is converted into heat[16]

$$P = V \cdot I$$

---

[14]Unit $[\kappa] \frac{\Omega \cdot \text{mm}^2}{\text{m}}$ in most case $\Omega \cdot$ m.

[15]The current flows from places with higher potential to places with lower potential.

[16]power = Energy/time = energy flow, unit of power: W (Watt).

The conductor has the resistance

$$R = \frac{V}{I}$$

Thus, the power P converted into heat in the resistor R[17]

$$P = I^2 \cdot R \tag{1.23}$$

Looking at the prism-shaped section in Fig. 1.11, the power dP converted in the prism into heat is obtained with (1.17) and (1.18) and analogous to (1.23)

$$dP = (dI)^2 \cdot \frac{dn}{\sigma \cdot dA} = J^2 \cdot (dA)^2 \cdot \frac{dn}{\sigma \cdot dA}$$

$$dP = \frac{1}{\sigma} \cdot J^2 \cdot dn \cdot dA \tag{1.24}$$

The product $dn \cdot dA$ is the elementary volume dV of the prism-shaped section of the field of current densities. For the power related to the elementary volume, i.e., the power density p

$$p = \frac{1}{\sigma} \cdot J^2 \tag{1.25}$$

With (1.21), the power density is[18]

$$p = J \cdot \left| \vec{E} \right| = \sigma \cdot E^2 \tag{1.26}$$

1.5   Electric Current in Metallic Conductors

The prerequisites for current in a metal conductor, e.g., in a copper conductor, are the free electrons of the metal atoms. Each atom in the atomic lattice of copper provides a free electron participating at high speed in the thermal chaos movement in the lattice. The quadratic mean value of this velocity is proportional to the temperature. The electrical current occurs as a result of the mechanical forces in the electrical field generated by the current or voltage source. This causes a directed movement of the electrons. In the current flow, the thermal chaos movement of the free electrons is superimposed by the slow, joint drift movement in the direction from the negative pole to the positive pole of an external voltage source. The free electrons are the carriers of the negative charge; the positively charged metal ions are the carriers of the positive charge.

---

[17]P is the power dissipation in the conductor.

[18]Unit $[p] = \frac{W}{m^3} = \frac{V \cdot A}{m^3}$.

The current is the quantity of charge that moves through the conductor cross section per unit of time. The unit of the quantity of charge is Coulomb (unit symbol C)[19]. A current of one Ampere[20] (unit symbol A) is defined as the quantity of electrical charge of 1 C (C) transported through the cross section of a conductor within one second.

Thus

$$1\text{A} = 1\frac{\text{C}}{\text{s}} \tag{1.27}$$

At a current density $J = 1\,\text{A/cm}^2$, a quantity of charge 1 C or 1 A · s is moving through a cross section of 1 cm². The charge of one electron is

$$e = -1{,}602\ 176\ 634 \cdot 10^{-19}\ \text{As} = -1{,}602\ 176\ 634 \cdot 10^{-19}\ \text{C}$$

Therefore, the quantity of charge of 1 C or of 1 A · s amounts to

$$\frac{1}{1{,}602\ 176\ 634 \cdot 10^{-19}} = 6{,}241\ 509\ 074 \cdot 10^{18}$$

electrons. A current I in a conductor is equivalent to the total charge Q, transported during the time t

$$Q = I \cdot t \tag{1.28}$$

At a drift velocity v and at a current I, a charge of

$$Q = I \cdot \frac{l}{v}$$

is transported in the conductor section of the length l. In a conductor cross section A, the current density J therefore is

$$J = \frac{I}{A} = \frac{Q \cdot v}{l \cdot A}$$

With (1.26) the power, converted into heat, in the conductor with the cross section A and the length l is

$$P = p \cdot l \cdot A = J \cdot E \cdot l \cdot A = \frac{Q \cdot v}{l \cdot A} \cdot E \cdot l \cdot A = Q \cdot v \cdot E \tag{1.29}$$

The free electrons of the conductor material move as a result of the force F exerted on them by the electric field. The mechanical power P required for this movement is

---

[19]Charles Augustin de Coulomb, French physicist and engineer, *1736, †1806.
[20]The definition of the unit of the current is explained in Sect. 3.2.1. Since the unit Ampere is a base unit, the unit Coulomb is a derived unit.

$$P = F \cdot v \tag{1.30}$$

The mechanical power P must be supplied to the electric field. The electric power P in (1.29) is equal to the mechanical power P in (1.30). Thus, the force F acting on the free electrons

$$\vec{F} \cdot \vec{v} = Q \cdot \vec{v} \cdot \vec{E}$$

and

$$\vec{F} = Q \cdot \vec{E} \tag{1.31}$$

Since the free electrons are negatively charged, the force exerted on the electrons acts against the direction of the electric field.

The integral of the force, along the path between two points a and b (path element ds) of the electric field, is equal to the energy W required to move the charge Q from point a to point b. According to (1.31)

$$W = \int_a^b \vec{F} \cdot d\vec{s} = Q \cdot \int_a^b \vec{E} \cdot d\vec{s} \tag{1.32}$$

The integral on the right side of this equation, according to (1.13), is the potential difference $(\varphi_a - \varphi_b)$, i.e., the voltage V_{ab} between the points a and b

$$\int_a^b \vec{E} \cdot d\vec{s} = -\int_b^a \vec{E} \cdot d\vec{s} = \varphi_b - \varphi_a = V_{ab} \tag{1.33}$$

This means that the voltage between two points a and b of the electric field describes the energy W required to move a charge $Q = 1C$ between these points. According to (1.6)

$$\vec{E} = -\mathrm{grad}\,\varphi = -\nabla\varphi$$

we obtain

$$W = Q \cdot V_{ab} = Q \cdot \int_a^b \vec{E} \cdot d\vec{s} = Q \cdot \int_a^b (-\nabla\varphi) \cdot d\vec{s} = -Q \cdot \int_a^b (\nabla\varphi) \cdot d\vec{s} \tag{1.34}$$

The gradient indicates the rise of the potential field. Since the electrons have a negative charge and $\varphi_b > \varphi_a$, according to (1.34) energy is required to move electrons from a lower potential level to a higher potential level, contrary to the direction of the electric field (see Fig. 1.5).[21]

---

[21]In physics, often the energy is referred to as eV.

Electrostatics

Electrostatics is the discipline of physics dealing with time-independent, i.e., static electric fields. Static electric fields are generated by static electric charges and charge distributions. Figure 2.1 shows an arrangement where a static electric field can be generated between two metallic electrodes. The arrangement is called capacitor. Between the electrodes, there is a non-conductive material, the dielectric[1]. The electrodes are connected to a voltage source, where negative charge carriers, i.e., electrons, are supplied to the electrode B. In electrode A, the free electrons of the metal are removed, resulting in positively charged metal ions, the carriers of the positive charge. If the terminals of the voltage source are disconnected from the electrodes, the charge on the electrodes is retained. The capacitor is therefore capable to store charges.

In the dielectric of the capacitor, an electric field is generated exerting a force on charges. The existence of the electric field can be proven by this force effect.

The force exerted on charges in the electric field can be measured by means of a test arrangement as shown in Fig. 2.2. The electrodes of the capacitor consist of two metal plates that are about 1 cm apart. The dielectric is air[2]. For the test, a high voltage source is used to apply a voltage of 15 kV to the electrodes of the capacitor. In the space between the electrodes is a small metal disk (charge spoon) which is connected to an electronic precision balance via a rod made of non-conductive material. The metal disk must first be unloaded. This is done with the voltage source switched off by touching the disk, whereby the experimenter has previously been earthed by one of the laboratory's earth lines or on a water line. The charge spoon is then put in the center of the dielectric,

---

[1]In the following, only non-conductive dielectrics are considered.

[2]The measured values for the case that the dielectric is air correspond approximately to the values for the case that there is a vacuum between the plates.

© Springer Fachmedien Wiesbaden GmbH, part of Springer Nature 2020
J. Donnevert, *Maxwell's Equations*, https://doi.org/10.1007/978-3-658-29376-5_2

Fig. 2.1 Capacitor

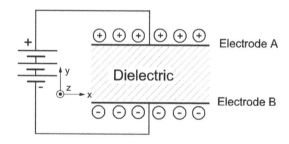

Fig. 2.2 Verification of the force in the electric field

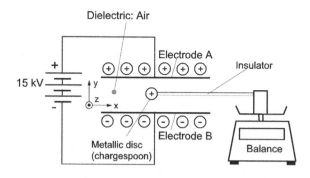

and the force exerted to the spoon is measured. If the voltage of 15 kV is applied to the electrodes and the deflection of the scale of the balance does not change, the charge spoon is in the correct position.

Two test series are carried out to measure the force excerted to the spoon. First, the charge spoon is positively charged by briefly bringing it into contact with the positively charged electrode. The charge on the spoon is then halved by touching it with a similar, uncharged spoon in a field-free space (voltage switched off). This procedure is repeated until the force on the measuring spoon is too small for a meaningful measurement.

Figure 2.3 shows the result of the measurement. We can see that the force F acting on a charge, at constant voltage V, is proportional to the amount of the charge Q

$$F \sim Q \quad \text{with } V = \text{const}$$

According to (1.31), the quotient F/Q is the magnitude of the electric field. The electric field is a vector pointing into the direction of the force acting on a positive charge

$$\vec{F} = Q \cdot \vec{E} \tag{2.1}$$

In a second series of tests, the force F acting on a charge Q is measured as a function of the voltage V applied to the electrodes of the capacitor. Figure 2.4 shows the measurement results. The force acting on the charge is proportional to the applied voltage.

$$F \sim V \quad \text{with } Q = \text{const}$$

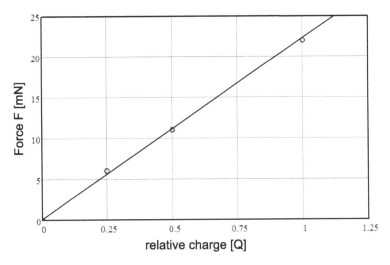

Fig. 2.3 Measurement of the force (A balance measures the gravitational force. Since the mean acceleration due to gravity is g $= 9.18$m$/$, s^2, the gravitational force $9.81\,N$ acts on an object of mass 1 kg. Conversely, 1 N (N) is the gravitational force which acts on an object with a mass of 102 grams.) F as a function of the charge ($V = 15\,$kV)

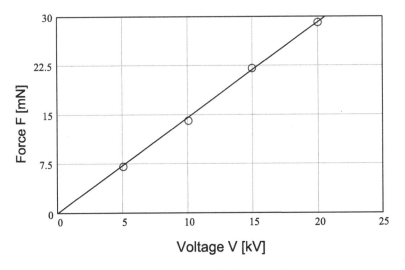

Fig. 2.4 Measurement of the force F as a function of voltage V at the capacitor

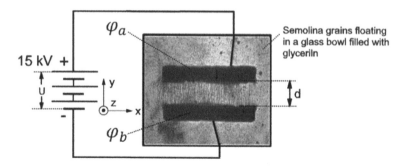

Fig. 2.5 Semolina picture of the electric field of a plate capacitor. (Reprint by courtesy of the Faculty of Physics of the LMU Munich)

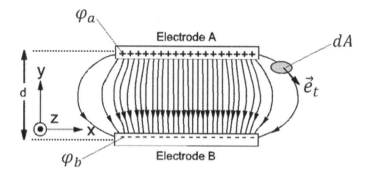

Fig. 2.6 Electric field lines of a plate capacitor (computer simulation)

With the help of semolina grains floating in a glass container filled with castor oil or glycerin, the field lines[3] of the electric field, formed between the electrodes of the capacitor, can be made visible. Castor oil, unlike tap water, is an insulator. Grains of semolina align themselves in an electric field along the direction of the field. Figure 2.5 shows the test arrangement. In the glass container, there are two electrodes that simulate the capacitor plates. The electrodes are connected to a high voltage source. The orientation of the semolina grains can be seen particularly well in the translucent light of a daylight projector. In addition to the semolina picture, a computer simulation of the electric field of a capacitor is shown in Fig. 2.6. The potential surfaces are not shown; they are perpendicular to the drawing plane.

---

[3]Field lines are lines that illustrate the directions of the vectors in vector fields. At each point of a field line, the tangent to the field line matches the direction of the vector in that field point.

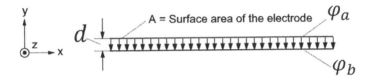

Fig. 2.7 Plate capacitor with small spacing *d* between the electrodes

Fig. 2.8 Measuring the charge
of a capacitor

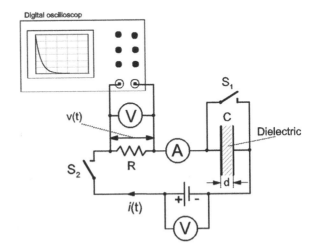

In technical applications, the distance *d* of the electrodes of a capacitor is always very small. The field lines outside the space between the electrodes are therefore negligible, and the field lines between the electrodes run parallel to each other (see Fig. 2.7). The electrodes are potential surfaces with the potentials φ_a and φ_b. The field between the potential surfaces is given by (see 1.1 and 1.4)

$$E = \frac{\varphi_a - \varphi_b}{d} = \frac{V}{d} \qquad (2.2)$$

The quantity of charge on an electrode of the capacitor, is called the electrical flux. At the end of the charging process the charge on the electrodes of the capacitor can be measured with the help of the arrangement according to Fig. 2.8.

At the beginning of the measurement, the switch S_1 is closed to discharge the capacitor C. After the capacitor has been discharged, the switch S_1 is opened and then the switch S_2 is closed. After closing the switch S_2, the charging current $i(t)$ flows. The amperemeter A is a sensitive measuring device (transient recorder) which measures the values of the instantaneous charging current at short time intervals and transmits them to a computer which calculates the integral of the current over the charging time.

Fig. 2.9 Charge
current of a capacitor
($V = 330\,\text{V}, d = 4\,\text{mm}$,
surface of the electrode:
$A = 900\,\text{cm}^2, R = 50\,\text{k}\Omega$,
dielectric: air)

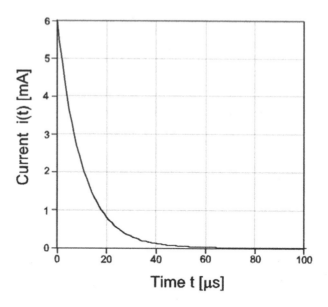

Alternatively, the voltage $v(t)$ at the resistor R can also be measured with a digital storage oscilloscope. From the time-varying voltage $v(t)$ at the resistor R the time-varying current $i(t)$ can be calculated and, by integration, the charge Q located on the electrodes of the capacitor at the end of the charging process (see 1.28)

$$Q = \int i(t)\mathrm{d}t = \frac{1}{R} \cdot \int u(t)\mathrm{d}t \tag{2.3}$$

Figure 2.9 shows the result of such a measurement. The result of the integration in this example is

$$Q = 5.98 \cdot 10^{-8}\,\text{As} = 5.98 \cdot 10^{-8}\,\text{C} \tag{2.4}$$

The electric field of a capacitor is illustrated in Fig. 2.6 by the field lines. The field lines start from the positive charges and end in the negative charges. The totality of the electric field lines, which are also referred to as lines of force or for historical reasons as displacement lines, form the so-called electric flux or displacement flux[4]. The totality of the electric field lines corresponds to the quantity of the charge on the electrodes. For both the electric flux and the charge producing it, the term Q is used. However, the electric flux is not to the same as the charge that causes it. The electric flux exists outside the charges. It represents the electric field generated by the charges. The electric field lines symbolize the force that the electric field exerts on charges in its sphere of influence. Electric field lines are comparable with current density lines, but in contrast to them

---

[4]The term displacement flux goes back to Maxwell, who described the electrical flux density as "displacement current".

there is no charge transport along the electric field lines[5]. The closer the electric field lines are to each other, the stronger is the field in this location. For the following considerations of the electric field, the electric flux density[6] is of particular importance. The electric flux density is a vector pointing in the direction of the tangent of the electric field line at the relevant field point. Figure 2.6 shows the surface element dA and the unit vector $\vec{e}_t$ which points into the direction of the tangent of the electric field line. The electrical flux dQ passes through the surface element dA. The electrical flux density $\vec{D}$ within the surface element dA is

$$\vec{D} = \frac{dQ}{dA} \cdot \vec{e}_t \tag{2.5}$$

The spacing d between the plates of a capacitor is always very small. Therefore, the electric charge and thus also the electrical flux Q can be assumed evenly distributed over the surface A of the electrodes, and the influence at the margins can be neglected. In this case, the electrical flux density can be calculated as follows:

$$\left|\vec{D}\right| = D = \frac{Q}{A} \tag{2.6}$$

With the following datas (see. Fig. 2.4 and the result of 2.4)

Voltage at the electrodes	$V = 300$ V
Distance between the plates	$d = 4$ mm
Surface of an electrode (plate)	$A = 900$ cm^2
Charge of the electrode	$Q = 5.976 \cdot 10^{-8}$ As

Using the data above (see Fig. 2.4 and the result of equation 2.4) we get the ratio D/E for the case that the dielectric is air or vacuum:

$$\frac{D}{E} = \frac{d}{V} \cdot \frac{Q}{A} = \frac{4 \cdot 10^{-3} \text{ m}}{300 \text{ V}} \cdot \frac{5.98 \cdot 10^{-8} \text{ As}}{900 \cdot 10^{-4} \text{ m}^2} = 8.859 \cdot 10^{-12} \frac{\text{As}}{\text{Vm}} \tag{2.7}$$

The charge measurement described in Fig. 2.8 can be performed with different insulating materials as dielectric at constant distance d between the plates (Results see Tab. 2.1)

---

[5]In the case of a non-conductive dielectric.

[6]Instead of the term electrical flux density, the term displacement density is also used for historical reasons. In the following text, only the term electrical flux density is used. It is the analog to the magnetic flux density (see Chap. 3).

Table 2.1 Relative permittivity values for different materials (18 °C, frequency > 100 Hz)

Dielectric	ε_r
Air	1.0006
Glass	2–3
Pertinax, epoxy resin	4.3–5.4
Ceramics	<10
Barium titanate	10^3–10^4

For free space (vacuum) between the electrodes of the capacitor, the ratio D/E is called electrical field constant ε_0, or absolute permittivity, or dielectric conductivity of the vacuum. The value of the absolute permittivity[7] is

$$\frac{D}{E} = \varepsilon_o = 8.85418782 \cdot 10^{-12} \frac{As}{Vm} \tag{2.8}$$

From (2.8), the electric field and the electrical flux density in a vacuum are related by

$$D = \varepsilon_o \cdot E \tag{2.9}$$

Since the vector of the electric flux density has the same direction as the electric field, we get

$$\vec{D} = \varepsilon_o \cdot \vec{E} \tag{2.10}$$

If the space between the electrodes is filled with a dielectric other than air, and the same voltage applied to the capacitor, larger amounts of charge can be stored on the capacitor, i.e., the electrical flux density will be greater than in the case of vacuum. The relationship between electric field and electric flux density in this case is determined by the product of the relative permittivity ε_r of the dielectric and the absolute permittivity ε_o

$$\vec{D} = \varepsilon_r \cdot \varepsilon_o \cdot \vec{E} = \varepsilon \cdot \vec{E} \tag{2.11}$$

Table 2.1 shows the relative permittivity of some materials. The increased storage capacity of capacitors with dielectric, whose relative permittivity is $\varepsilon_r > 1$, is based on the ability of the dielectric to align its molecules in the direction of the electric field. This orientation is called the polarization of the dielectric.

A capacitor is used to store a certain charge Q at a certain voltage V. If the voltage source is removed, the charge on the electrodes will be retained, provided that the dielectric is ideally non-conductive (specific conductivity equal to zero). From (2.11) follows with (2.2) for a plate capacitor with closely spaced plates

$$D = \frac{Q}{A} = \varepsilon \cdot E = \varepsilon \cdot \frac{V}{d}$$

---

[7]This value is derived in Chap. 4 from the magnetic field constant or the permeability of the vacuum μ_0 and the speed of light.

or

$$Q = \frac{\varepsilon \cdot A}{d} \cdot V \qquad (2.12)$$

In (2.12), the factor

$$C = \frac{\varepsilon \cdot A}{d} \qquad (2.13)$$

is the capacitance of the capacitor. From (2.12) and (2.13) follows

$$Q = C \cdot V \qquad (2.14)$$

The quantity of charge, a capacitor is capable to store, is proportional to the voltage applied to the capacitor.

2.1   Electric Field of Spherical and Point Charges

Electric fields do not only exist between the electrodes of a plate capacitor but also between charged objects of any shape. Next, electric fields between objects with spherical surfaces are examined in more detail.

In Fig. 2.10, a cross section of a metal sphere is shown. It is connected to the positive electrode of a DC voltage source so that a positive charge is applied to it. Also, the counter electrode can be assumed to be spherical concentric to the positive-charged metal sphere. It is located at an infinity far distance from the metal sphere. Since the negative electrode of the voltage source is grounded, the counter electrode is negatively charged.

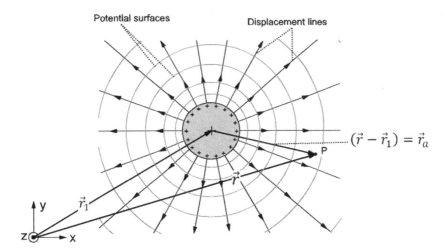

Fig. 2.10 Electric field lines and potential surfaces of a spherical charge (section in the surface at $z = 0$)

If the voltage source is removed, the charge will remain on the metal sphere and the counter electrode.

Since the charges on the surface of the sphere have the same sign, they repulse each other and thus are evenly distributed on the surface of the sphere. Under these conditions due to the symmetry, there is no potential difference on the surface of the sphere. In addition, the inside of the sphere is potential free. If an electric field was present inside the sphere, forces would be exerted on the free electrons of the metal that would cause the electrons to move. This would generate heat without the presence of an energy source. This contradicts the principle of conservation of energy. As a result, the surface of the sphere is a potential surface. If the sphere carries positive charges, the electric field lines emanate from the sphere. The magnitude of the electrical flux density $\vec{D}$ thus decreases with the square of the distance from the center of the sphere. By analogous application of (2.6) (charge Q of the sphere divided by the surface of the sphere), the following relationship is obtained for the vector of the electrical flux density $\vec{D}(\vec{r})$ outside the sphere (see Fig. 2.10):

$$\vec{D}(\vec{r}) = \frac{Q \cdot (\vec{r} - \vec{r}_1)}{4 \cdot \pi \cdot |\vec{r} - \vec{r}_1|^3} \tag{2.15}$$

In (2.15), the quotient

$$\frac{(\vec{r} - \vec{r}_1)}{|\vec{r} - \vec{r}_1|} = \vec{e}_{r\alpha}$$

is the unit vector showing into radial direction, i.e., it shows in the direction $(\vec{r} - \vec{r}_1) = \vec{r}_a$. With (2.11), the electric field is

$$\vec{E}(\vec{r}) = \frac{Q \cdot (\vec{r} - \vec{r}_1)}{4 \cdot \pi \cdot \varepsilon \cdot |\vec{r} - \vec{r}_1|^3} = \frac{Q}{4 \cdot \pi \cdot \varepsilon \cdot r_a^2} \cdot \vec{e}_{r\alpha} \tag{2.16}$$

As the electric field (see 1.5) is

$$\vec{E} = -\operatorname{grad} \varphi = -\nabla\varphi$$

and since the electric field has only a component in the radial direction, i.e., in the direction of $(\vec{r} - \vec{r}_1) = \vec{r}_a$ or $\left(\frac{\vec{r}_a}{r_a} = \vec{e}_{r\alpha}\right)$, with (1.9), we obtain

$$\vec{E}(\vec{r}_a) = -\frac{\partial\varphi}{\partial r_a} \cdot \vec{e}_{r\alpha} \tag{2.17}$$

From this equation follows with (2.16)

$$\varphi(\vec{r}_a) = -\int E(\vec{r}_{a\alpha}) \cdot \vec{e}_r \cdot \partial r_a = -\frac{Q}{4 \cdot \pi \cdot \varepsilon} \cdot \int \frac{1}{r_a^2} \cdot \partial r_a = \frac{Q}{4 \cdot \pi \cdot \varepsilon} \cdot \frac{1}{r_a} + \varphi_0$$

The summand φ_0 can be regarded as the potential in an infinite distance and may be set to zero.

Fig. 2.11 Superposition of the field of several point charges

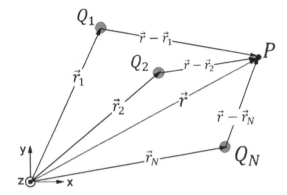

The potential outside the surface of the charged sphere is thus

$$\varphi(\vec{r}_a) = \varphi(\vec{r}) = \frac{Q}{4 \cdot \pi \cdot \varepsilon} \cdot \frac{1}{r_a} = \frac{Q}{4 \cdot \pi \cdot \varepsilon} \cdot \frac{1}{|\vec{r} - \vec{r}_1|} \tag{2.18}$$

The electric field outside the sphere in Fig. 2.10 can be assumed to be generated by a charge concentrated in the center of the sphere, i.e., by a point charge. If there are N point charges in space, the electric field of these charges results, since it is a linear system, from the sum of the electric flux densities or the electric field of the N individual charges (see Fig. 2.11).

For the electric flux density, the electric field and the potential in point P, with (2.15) to (2.18), we obtain

$$\vec{D}(\vec{r}) = \frac{1}{4 \cdot \pi} \sum_{j=1}^{N} \frac{Q_j \cdot (\vec{r} - \vec{r}_j)}{\cdot |\vec{r} - \vec{r}_j|^3} \tag{2.19}$$

respectively

$$\vec{E}(\vec{r}) = \frac{1}{4 \cdot \pi \cdot \varepsilon} \sum_{j=1}^{N} \frac{Q_j \cdot (\vec{r} - \vec{r}_j)}{\cdot |\vec{r} - \vec{r}_j|^3} \tag{2.20}$$

and respectively

$$\varphi(\vec{r}) = \frac{1}{4 \cdot \pi \cdot \varepsilon} \sum_{j=1}^{N} \frac{Q_j}{|\vec{r} - \vec{r}_j|} \tag{2.21}$$

Figure 2.12 shows as an example the semolina picture of an electric dipole. Ideally, the dipole consists of two point charges with opposite signs.

Figure 2.13 shows the calculated diagram of the electric field of an electric dipole. In addition to the electric field lines, the cross sections of two spherical surfaces A_1 and A_2 are shown in this diagram. These are closed enveloping surfaces, of which the

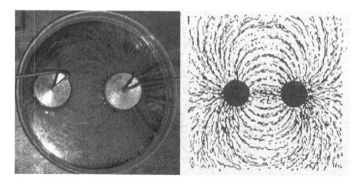

Fig. 2.12 Semolina picture of an electric dipole (*Source*: Joachim Herz Foundation)

enveloping surface A_2 encloses the charge $+Q$. The vector $\mathrm{d}\vec{A}$ is perpendicular to the respective surface element of the enveloping surface. As agreed, its direction is outward. Its magnitude is equal to the area of the surface element. If we calculate the scalar product $\vec{D} \cdot \mathrm{d}\vec{A}$, in one point of the enveloping surface, only the component of the electrical flux density $\vec{D}$ oriented perpendicular to the enveloping surface contributes the value to the product. The component of $\vec{D}$ which is tangential to the enveloping surface does not contribute to the scalar product $\vec{D} \cdot \mathrm{d}\vec{A}$. If the component of $\vec{D}$ is inward-looking, its contribution to the scalar product $\vec{D} \cdot \mathrm{d}\vec{A}$ will be negative.

In contrast to the surface A_2, the surface A_1 encloses neither of the two charges $+Q$ or $-Q$. Therefore, the electric flux that emerges from the surface A_1 is equal to the flux entering into the surface. The integral of the electrical flux density over the surface A_1 is therefore equal to zero:

$$\oiint_{A_1} \vec{D} \cdot \mathrm{d}\vec{A} = 0 \tag{2.22}$$

The surface A_2 encloses the charge $+Q$. From this surface, an electric flux emerges, of which the value is equal to the enclosed charge. The integral of electric flux density over the surface A_2 therefore has the value Q

$$\oiint_{A_2} \vec{D} \cdot \mathrm{d}\vec{A} = Q \tag{2.23}$$

A basic law of the electric field can be formulated according to (2.23) as follows:

In the electrostatic field, the integral of the electric flux density over any enveloping surface is equal to the charge enclosed by the enveloping surface.

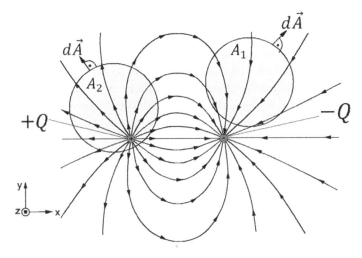

Fig. 2.13 Field lines of an electric dipole (section through the surface $z = 0$)

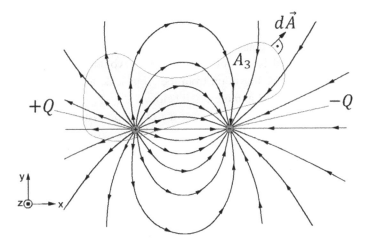

Fig. 2.14 Field lines of an electric dipole (section through the surface $z = 0$)

In Fig. 2.14, the image of the field of the electrical dipole shows a surface A_3 from which, as in the case of the surface A_1 in Fig. 2.13, electric field lines both enter and exit. However, in contrast to the enveloping surface A_1, it encloses the charge $+Q$. The difference between inflowing and outflowing electric flux and thus also the result of the integral

$$Q = \oiint\limits_{A_3} \vec{D} \cdot d\vec{A} \tag{2.24}$$

in this case is equal to $+Q$.

2.2   Spatial Charge Density

With (2.22) and (2.23), the electric field is considered on a quasi-large-scale basis. In a space where the charge is not concentrated in a few points in the form of spherical charges or point charges but is distributed in a certain form in the space V, a small-scale approach must be applied. In this case, the charge distribution is characterized by the spatial charge density ϱ at the respective space point, i.e., by the charge dQ related to the elementary volume dV (see Fig. 2.15):

$$\varrho = \frac{dQ}{dV} \tag{2.25}$$

The entire charge Q within the volume V is equal to the integral of the flux density $\vec{D}$ over the enveloping surface of this volume (see 2.23). Accordingly, the spatial charge density is the limit of the integral of the electrical flux density $\vec{D}$ over the enveloping surface A of the volume V divided by the volume V where the volume approaches to zero:

$$\varrho = \lim_{V \to 0} \frac{1}{V} \cdot \oiint_A \vec{D} \cdot d\vec{A} \tag{2.26}$$

In this equation, the volume considered is reduced to the elementary volume dV and the surface to be integrated is that of the elementary volume which is infinitesimal. Equation (2.26) represents the transition from the large-scale to the small-scale view of the electric field to be discussed in more detail below.

Using Fig. 2.16, the integral in (2.26) over the enveloping surface of the elementary volume dV is calculated. The result is the difference between the electric flux leaving the enveloping surface of the elementary volume dV and the flux entering it.

In the case of Cartesian coordinates, the elementary volume dV is a cube with the lengths dx, dy, and dz of its lateral surfaces. The components D_x, D_y, and D_z of the

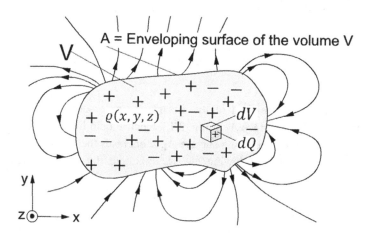

Fig. 2.15 Spatial charge density

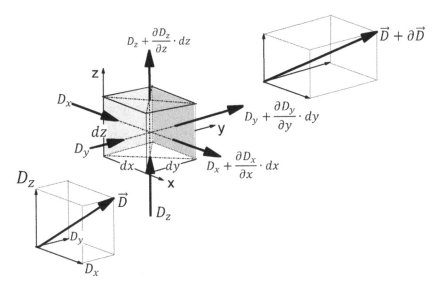

Fig. 2.16 Elementary volume $\mathbf{d}V$ for the evaluation of (2.26) in Cartesian coordinates

electrical flux density $\vec{D}$ are aligned perpendicular to the lateral surfaces of the elementary cube. Since the volume element is small, the component D_y on the surface $(\mathrm{d}x \cdot \mathrm{d}z)$ can be assumed to be constant over this surface. This is also true for the components D_x and D_z on the surface elements $(\mathrm{d}y \cdot \mathrm{d}z)$ and $(\mathrm{d}x \cdot \mathrm{d}y)$. The electrical flux that enters into the surface $(\mathrm{d}x \cdot \mathrm{d}z)$ is

$$D_y \cdot (\mathrm{d}x \cdot \mathrm{d}z)$$

The electric flux emerging from the opposite surface is

$$D_y \cdot (\mathrm{d}x \cdot \mathrm{d}z) + \frac{\partial \left(D_y \cdot \mathrm{d}z \cdot \mathrm{d}x\right)}{\partial y} \cdot \mathrm{d}y$$

The variation of the flux in y-direction results as

$$\frac{\partial D_y}{\partial y} \cdot \mathrm{d}y \cdot \mathrm{d}z \cdot \mathrm{d}x$$

The variation of the flux in x-direction respectively z-direction results in

$$\frac{\partial D_x}{\partial x} \cdot \mathrm{d}x \cdot \mathrm{d}y \cdot \mathrm{d}z$$

and

$$\frac{\partial D_z}{\partial z} \cdot \mathrm{d}z \cdot \mathrm{d}y \cdot \mathrm{d}x$$

The limit in (2.26) is equal to the flux leaving the elementary volume minus the flux entering the surfaces of the elementary volume divided by the elementary volume dV, i.e., the flux variation, divided by the elementary volume dV

$$dV = dx \cdot dy \cdot dz$$

The charge density ϱ present in the volume element dV is therefore

$$\varrho = \lim_{V \to 0} \frac{1}{V} \cdot \oiint_A \vec{D} \cdot d\vec{A} = \frac{\partial D_x}{\partial x} + \frac{\partial D_y}{\partial y} + \frac{\partial D_z}{\partial z} \tag{2.27}$$

The integral and thus also the sum in (2.27) is called divergence[8] of the electrical flux density $\vec{D}$ and is abbreviated by the symbol div

$$\operatorname{div} \vec{D} = \lim_{V \to 0} \frac{1}{V} \cdot \oiint_A \vec{D} \cdot d\vec{A} = \frac{\partial D_x}{\partial x} + \frac{\partial D_y}{\partial y} + \frac{\partial D_z}{\partial z} = \varrho \tag{2.28}$$

Instead of the designation div in (2.28), the Nabla operator ∇ is usually used

$$\operatorname{div} \vec{D} = \nabla \cdot \vec{D} = \varrho \tag{2.29}$$

Statement of this equation

The vector operator $(\nabla \cdot)$ (pronounced: "del dot") applied to the vector field of the electric flux density $\vec{D}$ indicates that the spatial variation of the flux density in the elementary volume is equal to the charge density ϱ present at the point in space under consideration.

The operator ∇ applied to a vector field has the following form in Cartesian coordinates (see 1.7):

$$\nabla = \frac{\partial}{\partial x} \cdot \vec{e}_x + \frac{\partial}{\partial y} \cdot \vec{e}_y + \frac{\partial}{\partial z} \cdot \vec{e}_x \tag{2.30}$$

With the notation of (2.30)

$$\operatorname{div} \vec{D} = \nabla \cdot \vec{D} = \left(\frac{\partial}{\partial x} \cdot \vec{e}_x + \frac{\partial}{\partial y} \cdot \vec{e}_y + \frac{\partial}{\partial z} \cdot \vec{e}_x \right) \cdot \left(D_x \cdot \vec{e}_x + D_y \cdot \vec{e}_y + D_z \cdot \vec{e}_x \right) = \varrho \tag{2.31}$$

The charge in the elementary volume dV is $(\varrho \cdot dV)$. This charge is the source of the electric flux that flows out of the elementary volume. If the volume is free of sources, no electrical flux could emanate from it

$$\nabla \cdot \vec{D} = 0 \tag{2.32}$$

From the definition of the spatial charge density ϱ in (2.25), we obtain with (2.29) the charge Q which is present in a volume V

---

[8]latin divergere = to strive apart.

$$Q = \iiint\limits_V \varrho \cdot dV = \iiint\limits_V \left(\nabla \cdot \vec{D} \right) \cdot dV \tag{2.33}$$

The triple integral sign indicates that the integration is to be executed in the three-dimensional space. With (2.23)

$$\oiint\limits_A \vec{D} \cdot d\vec{A} = Q \tag{2.34}$$

The surface A is the closed enveloping surface of a volume V. From both Eqs. (2.33) and (2.34) follows Gauß's integral theorem[9]

$$\iiint\limits_V \left(\nabla \cdot \vec{D} \right) \cdot dV = \oiint\limits_A \vec{D} \cdot d\vec{A} \tag{2.35}$$

Equation (2.35) states:

> The volume integral of the divergence of a vector field, executed over the space V, is equal to the surface integral of this vector field executed over the enveloping surface A of this space. Note that the normal vector of the enveloping surface A is orientated towards the outside.

Following (2.21), the potential of a spatial charge that is continuously distributed in the space V with the spatial charge density $\varrho(\vec{r}_v)$ (see Fig. 2.17) is

$$\varphi(\vec{r}) = \frac{1}{4 \cdot \pi \cdot \varepsilon} \iiint\limits_V \varrho(\vec{r}_v) \cdot \frac{dV}{|\vec{r} - \vec{r}_v|} \tag{2.36}$$

The potential $\varphi(\vec{r})$ in (2.36) is called electric scalar potential.

Fig. 2.17 Spatial charge continuously distributed in the volume V with the spatial charge density $\varrho(\vec{r}_v)$

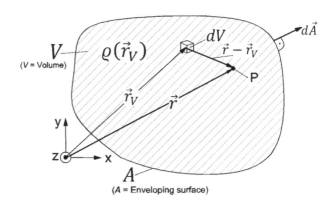

[9]Gauß, Karl Friedrich, German mathematician, astronomer, geodesist, and physicist, *1777, †1855.

2.2.1 Divergence div $\vec{D} = \nabla \cdot \vec{D}$ in Cylindrical and Spherical Coordinates

For cylinder-symmetric and spherical-symmetric problems, it is useful to perform the operation (div $= \nabla \cdot$) in cylindrical or spherical coordinates. Next, the operation (div $= \nabla \cdot$) for these coordinate systems will be derived.

Figure 2.18 shows the elementary volume dV in cylinder coordinates. The electrical flux density $\vec{D}$ in this case consists of the components D_z, D_r, and D_α. The elementary volume dV is created when the coordinates r, α, and z increase by dr, dα, and dz. It is permissible to assume the elementary volume as a cuboid. It can be seen from Fig. 2.18 that in this case the elementary volume dV consists of the lateral lengths dr, dz, and $(r \cdot d\alpha)$. Its volume is therefore

$$dV = dr \cdot (r \cdot d\alpha) \cdot dz$$

For the surface of the elementary volume (see Fig. 2.18), we have

Lateral area	Surface of the lateral area
$\alpha = $ const	$dr \cdot dz$
$r = $ const	$(r \cdot d\alpha) \cdot dz$
$z = $ const	$(r \cdot d\alpha) \cdot dr$

Variation of the electrical flux in z-direction

$$\frac{\partial \left[D_z \cdot dr \cdot (r \cdot d\alpha) \right]}{\partial z} \cdot dz = \frac{\partial D_z}{\partial z} \cdot (dr \cdot r \cdot d\alpha \cdot dz)$$

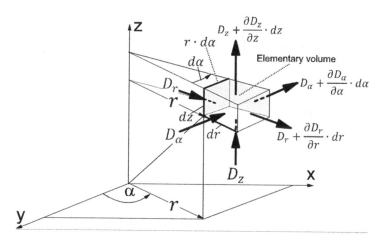

Fig. 2.18 Elementary volume in cylindrical coordinates

Variation of the electrical flux in r-direction

$$\frac{\partial[D_r \cdot dz \cdot (r \cdot d\alpha)]}{\partial r} \cdot dr = \frac{\partial(D_r \cdot r)}{\partial r} \cdot dr \cdot dz \cdot d\alpha = \frac{1}{r} \cdot \frac{\partial(D_r \cdot r)}{\partial r} \cdot (dr \cdot dz \cdot r \cdot d\alpha)$$

Variation of the electrical flux in α-direction

$$\frac{\partial(D_\alpha \cdot dz \cdot dr)}{\partial(\alpha \cdot r)} \cdot (r \cdot d\alpha) = \frac{\partial D_\alpha}{\partial(r \cdot \alpha)} \cdot r \cdot d\alpha \cdot dz \cdot dr = \frac{1}{r} \frac{\partial D_\alpha}{\partial \alpha} \cdot (r \cdot d\alpha \cdot dz \cdot dr)$$

The limit in (2.26) is the sum of the variation of the electrical flux divided by volume $dV = [dr \cdot (r \cdot d\alpha) \cdot dz]$. Thus the divergence of the flux density $\vec{D}$ is for cylinder coordinates

$$\nabla \cdot \vec{D} = \lim_{V \to 0} \frac{1}{V} \oiint_A \vec{D} \cdot d\vec{A} = \frac{\partial D_z}{\partial z} + \frac{1}{r} \cdot \frac{\partial(D_r \cdot r)}{\partial r} + \frac{1}{r} \frac{\partial D_\alpha}{\partial \alpha} \tag{2.37}$$

Figure 2.19 shows the elementary volume in spherical coordinates. The electrical flux density $\vec{D}$, in this case, consists of the components D_r, D_ϑ, and D_α. The elementary volume dV is created when the coordinates r, α, and ϑ increase by dr, $d\alpha$, and $d\vartheta$. The elementary volume dV has the following three lateral lengths in the case of the spherical coordinates (see Fig. 2.19):

$$dr, (r \cdot d\vartheta) \text{ and } (r \cdot \sin \vartheta \cdot d\alpha).$$

The elementary volume dV is

$$dV = [dr \cdot (r \cdot d\vartheta) \cdot (r \cdot \sin \vartheta \cdot d\alpha)]$$

The areas of the lateral surfaces of the elementary volume dV are

$$[dr \cdot (r \cdot d\vartheta)], [(r \cdot d\vartheta) \cdot (r \cdot \sin \vartheta \cdot d\alpha)] \text{ and } [dr \cdot (r \cdot \sin \vartheta \cdot d\alpha)].$$

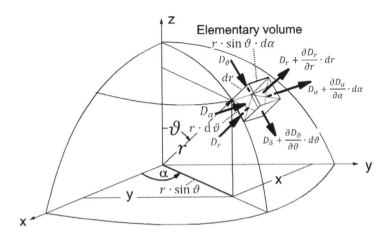

Fig. 2.19 Elementary volume in spherical coordinates

The variation of the electric flux in r-direction is therefore

$$\frac{\partial[D_r \cdot (r \cdot d\vartheta) \cdot (r \cdot \sin \vartheta \cdot d\alpha)]}{\partial r} \cdot dr = \frac{1}{r^2} \cdot \frac{\partial (D_r \cdot r^2)}{\partial r} \cdot [dr \cdot (r \cdot d\vartheta) \cdot (r \cdot \sin \vartheta \cdot d\alpha)]$$

Variation of the flux in ϑ-direction

$$\frac{\partial[D_\vartheta \cdot (dr) \cdot (r \cdot \sin \vartheta \, d\alpha)]}{r \cdot \partial \vartheta} \cdot (r \cdot d\vartheta)$$

$$= \frac{1}{r \cdot \sin \vartheta} \cdot \frac{\partial (D_\vartheta \cdot \sin \vartheta)}{\partial \vartheta} \cdot [(d\vartheta \cdot r) \cdot (dr) \cdot (r \cdot \sin \vartheta \cdot d\alpha)]$$

Variation of the flux in α-direction

$$\frac{\partial[D_\alpha \cdot dr \cdot (r \cdot d\vartheta)]}{\partial \alpha \cdot (r \cdot \sin \vartheta)} \cdot (r \cdot \sin \vartheta \cdot d\alpha) = \frac{1}{r \cdot \sin \vartheta} \frac{\partial D_\alpha}{\partial \alpha} \cdot [(d\vartheta \cdot r) \cdot dr \cdot (r \cdot \sin \vartheta \cdot d\alpha)]$$

The limit in (2.26) is the sum of the variations of the electrical flux divided by the volume dV. Vector operation divergence in the spherical coordinate system

$$\nabla \cdot \vec{D} = \lim_{V \to 0} \frac{1}{V} \cdot \oiint_A \vec{D} \cdot d\vec{A} = \frac{1}{r^2} \cdot \frac{\partial (r^2 \cdot D_r)}{\partial r} + \frac{1}{r \cdot \sin \vartheta} \cdot \frac{\partial (D_\vartheta \cdot \sin \vartheta)}{\partial \vartheta} + \frac{1}{r \cdot \sin \vartheta} \frac{\partial D_\alpha}{\partial \alpha}$$

$$(2.38)$$

2.3   Potential Equation of the Electrical Scalar Potential

According to (2.32) in a space without charge (space charge density $\varrho = 0$),

$$\nabla \cdot \vec{D} = \nabla \cdot \left(\varepsilon \cdot \vec{E} \right) = 0$$

and consequently

$$\nabla \cdot \vec{E} = 0 \qquad (2.39)$$

According to Eq. (1.5),

$$\vec{E} = -\nabla \varphi \qquad (2.40)$$

from both equations follows

$$\nabla \cdot \vec{E} = \nabla \cdot (-\nabla \varphi) = 0 \qquad (2.41)$$

respectively

$$\nabla \cdot \nabla \varphi = \nabla^2 \varphi = 0 \qquad (2.42)$$

Equation (2.42) is called the Laplace equation[10]. It is the potential equation of the electrical scalar potential for a space without electrical charge. In this equation, instead of the designation $\nabla \cdot \nabla \varphi$, the sign $\nabla^2 \varphi$ is introduced[11].

Since the gradient of a scalar field is a vector function and the divergence of a vector function results in a scalar function, the result of the vector operation $\nabla \cdot \nabla \varphi = \nabla^2 \varphi$ is a scalar.

$\nabla^2 \varphi$ in Cartesian Coordinates

With (1.7) and (2.30), we get

$$\nabla^2 \varphi = \left(\frac{\partial}{\partial x} \cdot \vec{e}_x + \frac{\partial}{\partial y} \cdot \vec{e}_y + \frac{\partial}{\partial z} \cdot \vec{e}_z \right) \cdot \left(\frac{\partial \varphi}{\partial x} \cdot \vec{e}_x + \frac{\partial \varphi}{\partial y} \cdot \vec{e}_y + \frac{\partial \varphi}{\partial z} \cdot \vec{e}_z \right)$$

Thus

$$\nabla^2 \varphi = \frac{d^2 \varphi}{\partial x^2} + \frac{d^2 \varphi}{\partial y^2} + \frac{d^2 \varphi}{\partial z^2} \tag{2.43}$$

$\nabla^2 \varphi$ in cylinder Coordinates

With (1.8) and (2.37), we have

$$\nabla \cdot \nabla \varphi = \nabla^2 \varphi = \frac{1}{r} \cdot \frac{\partial}{\partial r} \left(r \cdot \frac{\partial \varphi}{\partial r} \right) + \frac{1}{r} \frac{\partial}{\partial \alpha} \left(\frac{1}{r} \cdot \frac{\partial \varphi}{\partial \alpha} \right) + \frac{\partial}{\partial z} \left(\frac{\partial \varphi}{\partial z} \right)$$

With the product rule

$$\nabla^2 \varphi = \frac{d^2 \varphi}{\partial r^2} + \frac{1}{r} \cdot \frac{\partial \varphi}{\partial r} + \frac{1}{r^2} \frac{\partial^2 \varphi}{\partial \alpha^2} + \frac{\partial^2 \varphi}{\partial z^2} \tag{2.44}$$

$\nabla^2 \varphi$ in Spherical Coordinates

With (1.9), we get

$$\nabla \cdot \nabla \varphi = \nabla \cdot \left(\frac{\partial \varphi}{\partial r} + \frac{1}{r} \cdot \frac{\partial \varphi}{\partial \vartheta} + \frac{1}{r \cdot \sin \vartheta} \cdot \frac{\partial \varphi}{\partial \alpha} \right)$$

---

[10]Laplace, Pierre-Simon, French mathematician, physicist, and astronomer, *1749, †1827.

[11]In some books, the formula symbol Δ may be used for the vector operation ∇^2. It is referred to as the Delta or Laplace operator.

According to (2.38)

$$\nabla \cdot \nabla\varphi = \frac{1}{r^2} \cdot \frac{\partial}{\partial r}\left(r^2 \cdot \frac{\partial\varphi}{\partial r}\right) + \frac{1}{r \cdot \sin\vartheta} \cdot \frac{\partial}{\partial\vartheta}\left(\frac{1}{r} \cdot \frac{\partial\varphi}{\partial\vartheta} \cdot \sin\vartheta\right)$$

$$+ \frac{1}{r \cdot \sin\vartheta} \cdot \frac{\partial}{\partial\alpha}\left(\frac{1}{r \cdot \sin\vartheta} \cdot \frac{\partial\varphi}{\partial\alpha}\right)$$

Thus

$$\nabla^2\varphi = \frac{1}{r^2} \cdot \frac{\partial}{\partial r}\left(r^2 \cdot \frac{\partial\varphi}{\partial r}\right) + \frac{1}{r^2 \cdot \sin\vartheta} \cdot \frac{\partial}{\partial\vartheta}\left(\sin\vartheta \cdot \frac{\partial\varphi}{\partial\vartheta}\right) + \frac{1}{r^2 \cdot (\sin\vartheta)^2} \cdot \frac{\partial^2\varphi}{\partial\alpha^2}$$

$$(2.45)$$

Equation (2.29) refers to a space with the spatial charge

$$\nabla \cdot \vec{D} = \nabla \cdot \left(\varepsilon \cdot \vec{E}\right) = \varrho \tag{2.46}$$

According to (1.5), describing the relationship between potential and electric field

$$\vec{E} = -\nabla\varphi$$

and with (2.46), we have

$$\nabla \cdot \vec{E} = -\nabla \cdot (\nabla\varphi) = \frac{\varrho}{\varepsilon} \tag{2.47}$$

or

$$\nabla^2\varphi = -\frac{\varrho}{\varepsilon} \tag{2.48}$$

Equation (2.48) is the potential equation for a space with spatial charge. It is called the Poisson equation[12]. Equation (2.36) is the solution to this equation.

Potentials are functions derived from the vector field functions. In the present case, the electric field is the original field function of the electric scalar potential. Potentials are linked to the field functions via spatial integration. The field function is obtained by differentiating the potential function (see also 1.5 or 2.40). In this volume, the potential functions are steps on the way to Maxwell's equations.

---

[12]Poisson, Siméon Denis, French physicist and mathematician, *1781, †1840.

2.4   Energy Density of the Electric Field

On the electrodes of a capacitor of capacitance C, between which a voltage V is present, according to (2.14), a charge Q is stored

$$Q = C \cdot V$$

During the charging process, a charge flows from the current source to the electrodes of the capacitor; taking a certain amount of time, energy is transferred from the current source to the capacitor. At any time t, there is an instantaneous charge q on the electrodes. For this instantaneous charge, we have ($v =$ instantaneous voltage)

$$q = C \cdot v \tag{2.49}$$

The charge on electrodes of the capacitor is caused by a current i from the current source. The current i charges the capacitor with dq during the time dt

$$dq = i \cdot dt \tag{2.50}$$

According to Eq. (2.49), the voltage at the capacitor increases therefore by dv

$$dq = C \cdot dv \tag{2.51}$$

From both equations, we get

$$i \cdot dt = C \cdot dv \tag{2.52}$$

The instantaneous power consumed during the charging process is equal to the instantaneous current i multiplied by the instantaneous voltage v. Consequently, the energy dW absorbed during the charging process is

$$dW = v \cdot i \cdot dt = v \cdot C \cdot dv \tag{2.53}$$

The voltage of the capacitor at the end of the charging process is V. If the capacitor has no voltage at the beginning of the charging process, it has absorbed the energy W during the charging process

$$W = C \cdot \int_0^V v \cdot dv = \frac{1}{2} \cdot C \cdot V^2 \tag{2.54}$$

The energy absorbed by the capacitor is stored in the electric field of the capacitor. In order to determine the energy density of the electric field, a small volume element of the electric field is considered in the form of an elementary capacitor in which the electric field can be considered to be constant (see Fig. 2.20). It may be assumed that the front surfaces of the elementary capacitor are covered with thin metal foils. The capacitance C_{el} of the elementary capacitor is (see 2.13)

$$C_{el} = \varepsilon \cdot \frac{dA}{dn} \tag{2.55}$$

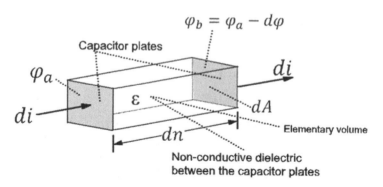

Fig. 2.20 Elementary capacitor

Between the two electrodes of the elementary capacitor exists the potential difference

$$\varphi_a - \varphi_b = \mathrm{d}\varphi$$

The voltage applied to the elementary capacitor is $\mathrm{d}\varphi$. The electric field between the electrodes of the elementary capacitor (see 2.2) is

$$E = \frac{\varphi_a - \varphi_b}{\mathrm{d}n} = \frac{\mathrm{d}\varphi}{\mathrm{d}n}$$

Therefore

$$\mathrm{d}\varphi = E \cdot \mathrm{d}n$$

Thus the energy $\mathrm{d}W$, stored by the elementary capacitor with the capacitance C_{el}, is to be calculated according to (2.54) and (2.56)

$$\mathrm{d}W = \frac{1}{2} \cdot C_{\mathrm{el}} \cdot (\mathrm{d}\varphi)^2 = \frac{1}{2} \cdot \varepsilon \cdot \frac{\mathrm{d}A}{\mathrm{d}n} \cdot (\mathrm{d}\varphi)^2 = \frac{1}{2} \cdot \varepsilon \cdot \frac{\mathrm{d}A}{\mathrm{d}n} \cdot E^2 \cdot (\mathrm{d}n^2)$$

$$\mathrm{d}W = \frac{1}{2} \cdot \varepsilon \cdot E^2 \cdot \mathrm{d}A \cdot \mathrm{d}n \tag{2.56}$$

In (2.56)

$$\mathrm{d}A \cdot \mathrm{d}n = \mathrm{d}V$$

is the volume of the elementary capacitor. For the energy stored in the volume element $\mathrm{d}V$, we have

$$\mathrm{d}W = \frac{1}{2} \cdot \varepsilon \cdot E^2 \cdot \mathrm{d}V \tag{2.57}$$

Finally, the energy density w_{el} stored in the electric field is

$$w_{\mathrm{el}} = \frac{\mathrm{d}W}{\mathrm{d}V} = \frac{1}{2} \cdot \varepsilon \cdot E^2 = \frac{1}{2} \cdot D \cdot E \tag{2.58}$$

The Stationary Magnetic Field

3

Flowing electric current, or moving electric charges, creates a magnetic field. An electric current, constant with time, generates a stationary magnetic field. In the simplest case, the current-carrying conductor is a straight copper wire. The magnetic field can be made visible with iron filings. Figure 3.1 shows an appropriate test arrangement. A copper conductor connected to a DC voltage source is passed through a glass plate sprinkled with iron filings.

When the switch S is closed, current flows and the iron filings align themselves in the direction of the magnetic field lines similar to small compass needles (see Fig. 3.2). The cause of the magnetic field is the current flowing through the conductor. The field lines illustrate the force that the magnetic field exerts on the iron filings. The tangent to a field line indicates the direction of the force acting on an iron particle. In this example, the magnetic field lines are concentric circles around the current-carrying conductor. The direction of the current and the direction of the magnetic field form a right-hand screw[1]. In contrast to the electric field lines emanating from positive charges and ending in negative charges, there are no sources and sinks of the magnetic field lines in the magnetic field.

---

[1] Direction of the magnetic field lines   Direction of the current

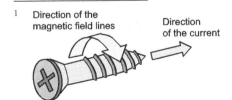

© Springer Fachmedien Wiesbaden GmbH, part of Springer Nature 2020   43
J. Donnevert, *Maxwell´s Equations*, https://doi.org/10.1007/978-3-658-29376-5_3

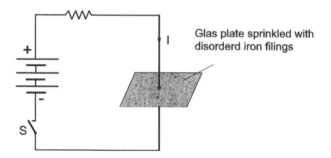

Fig. 3.1 Experimental arrangement with iron filings

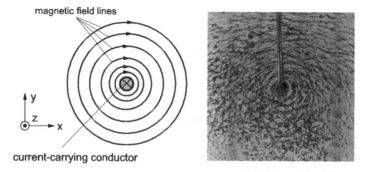

Fig. 3.2 Magnetic field lines of a current-carrying straight conductor (*left*: magnetic field lines forming a right-hand screw with the current entering the drawing plane, *right*: picture with iron filings, *source* Joachim Herz Foundation)

The intensity of the magnetic field is indicated by the density of the field lines. As in the case of the electric field, the density of the field lines is characterized by the flux density in this case by the magnetic flux density[2], designated by the vector $\vec{B}$. The magnitude of the magnetic flux density is proportional to the intensity of the current causing the magnetic field. The unit of magnetic flux density and the method for measuring it are explained in more detail below.

Figure 3.3 shows an inductor[3] with its current and the image of the magnetic field. As in the case of the straight conductor, the magnetic field lines are also closed. Outside the inductor, the magnetic lines are only partially shown. Inside the inductor, the field lines are almost parallel, and the magnetic field is almost homogeneous.

---

[2]The magnetic flux density is also called magnetic induction.

[3]Also called coil, choke, reactor.

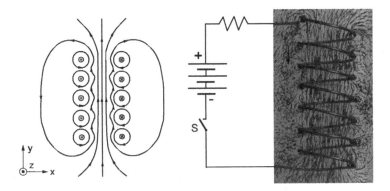

Fig. 3.3 Magnetic field lines of a current-carrying inductor and associated picture with iron filings (*Source* Joachim Herz Foundation)

3.1   Force Effect in the Stationary Magnetic Field

The experiments with iron filings show that a current through a conductor generates a magnetic field and that force is exerted on the iron filings in this magnetic field. On the other hand, force is also exerted vice versa on a conductor with current being in a magnetic field. This force can be verified and measured with a test arrangement as shown in Fig. 3.4. In this experimental setup, the magnetic field is generated by an excitation inductor. The windings of this inductor are mounted on two tubular bobbins. The windings, contrary to what is indicated in the figure, are close to each other and the windings of the two parts of the excitation inductor are connected to each other. The magnetic field

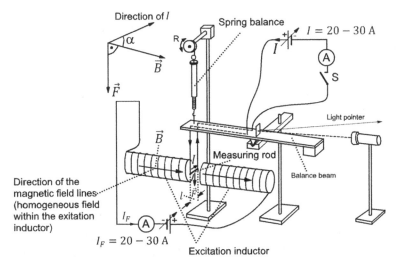

Fig. 3.4 Measurement of the force effect in the magnetic field (*Source* Joachim Herz Foundation)

inside the excitation inductor can be assumed to be homogeneous. In order to introduce a conductor loop with a measuring rod into the magnetic field inside the inductor, the two parts of the excitation inductor are separated by a narrow air gap. In the figure, the air gap had to be drawn larger in order to be able to show the suspension of the measuring rod. The measuring rod is horizontal and initially aligned at a right angle to the inductor axis.

The force acting on the measuring rod is measured by a spring balance. During the measurement, the position of the measuring rod in the magnetic field must not change. For this purpose, the measuring rod is attached to a balance beam. When the switch S is open, the balance is brought into equilibrium, whereby the spring balance is only slightly loaded. The balance beam should support the entire weight of the measuring rod, including its suspension. A light pointer indicates the position of the balance beam on a projection screen.

When the switch S is closed, current will flow through the conductor integrated in the balance beam, the suspension, and the measuring rod. As a result, a force acts on the measuring rod. With the current I in the direction drawn in the figure, this force acts downward. By turning wheel R, the balance beam is returned to its original position. The force acting on the measuring rod can then be read on the scale of the spring balance. The magnitude of the force in the test arrangement lies in the mN-range. During the following measurements, both the excitation current I_F of the excitation inductor and the current I through the measuring rod will be changed. Besides the two currents I and I_F, the angle α of the measuring rod is changed keeping a horizontal position.

The intensity of the excitation current I_F in the windings of the excitation inductor determines the intensity of the magnetic field. The vector $\vec{B}$ in Fig. 3.4 indicates the direction of the magnetic field lines within the excitation inductor and by its magnitude the intensity of the magnetic field

Results of the measurement with the current balance:

1. The magnitude of the force $\vec{F}$ acting on the measuring rod is proportional to the current I through the measuring rod. If the direction of the current in the measuring rod is reversed, the direction of the force will be reversed also.
2. The magnitude of the force $\vec{F}$ is proportional to the current I_F through the excitation inductor and thus to the density of the field lines, and thus to the intensity of the magnetic flux density $\vec{B}$ inside the excitation inductor.
3. The magnitude of the force $\vec{F}$ depends on the direction of the measuring rod in relation to the direction of the field lines. The greatest force is exerted when the rod is perpendicular to the magnetic field lines as shown in Fig. 3.4, i.e., for $\alpha = 90°$.
4. The force vector $\vec{F}$ is perpendicular to the surface defined by the direction of the current I through the measuring rod and the vector $\vec{B}$. The direction of the current I, the vector $\vec{B}$, and the vector $\vec{F}$ form a right-hand screw (see Fig. 3.5).
   The three-finger rule of the right hand can serve as a memory aid.
5. The magnitude $\left|\vec{F}\right|$ of the force is proportional to the length l of the measuring rod in the magnetic field.

Fig. 3.5 Three-finger rule of the right hand

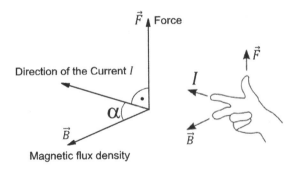

From these results, the following relationship can be derived for a homogeneous magnetic field in the region of the current-carrying conductor (measuring rod)

$$\left|\vec{F}\right| = F = I \cdot \left|\vec{B}\right| \cdot l \cdot \sin \alpha = B \cdot I \cdot l \cdot \sin \alpha \tag{3.1}$$

In (3.1), the magnetic flux density $\left|\vec{B}\right|$ is a proportionality factor. The greatest force $F = F_{max}$ is exerted when the rod is perpendicular to the magnetic field lines, i.e., when $\alpha = 90°$. Now we have

$$\left|\vec{B}\right| = B = \frac{F_{max}}{I \cdot l} \tag{3.2}$$

With (3.1) and (3.2), the magnetic flux density $B = \left|\vec{B}\right|$ is defined.

If a current of 1 A flows through a measuring rod of 1 m length located in and oriented perpendicular to the direction of a homogeneous magnetic field and if a force of 1 N is exerted on this rod by the magnetic field, then the magnetic flux density B will be 1 T (T = Tesla)[4]

$$B = 1 \frac{N}{A \cdot m} = 1\,T \tag{3.3}$$

Thus, the unit of the magnetic flux density is

$$\text{Unit}(B) = \frac{N}{A \cdot m} = \frac{s}{A \cdot m^2} \cdot N \cdot \frac{m}{s} = \frac{s \cdot W}{A \cdot m^2} = \frac{s \cdot A \cdot V}{A \cdot m^2} = \frac{V \cdot s}{m^2} = 1\,T \tag{3.4}$$

---

[4]T = Tesla, Nicola, Croatian-American electrical engineer and physicist, *1856, †1943.

The magnetic flux density $\vec{B}$ is a vector pointing into the direction of the magnetic field lines. The density of the field lines indicates the magnitude of $\left|\vec{B}\right|$. The total number of magnetic field lines, passing a surface A, is obtained by the integral of the vector $\vec{B}$ over this surface. The magnetic flux Φ is the integral over the flux density $\vec{B}$.

$$\Phi = \iint\limits_A \vec{B} \cdot d\vec{A}$$

(3.5)

Unit of the magnetic flux

$$\text{Unit}(\Phi) = \frac{V \cdot s}{m^2} \cdot m^2 = V \cdot s = 1\,Wb = 1\,Weber$$

(3.6)

[5]Since the magnetic field lines are closed, just as many field lines exit from a closed enveloping surface as field lines enter into this closed enveloping surface. For this reason, the integral of the magnetic flux over a closed surface A is zero

$$\oiint\limits_A d\vec{B} = 0$$

(3.7)

Equation (3.7) says that the magnetic field lines do not emanate from a source. There are no magnetic sources in the magnetic field. This is expressed in the following equation (cf. (2.32)):

$$\nabla \cdot \vec{B} = 0$$

(3.8)

The force effect of the magnetic field on a current-carrying conductor is known as the Lorentz[6] force. The Lorentz force is the force acting on moving charges. When current flows in a conductor, the Lorentz force acts on the free electrons moving in the conductor. A current I is equal to the quantity of charge Q which flows during time t (cf. (1.28))

$$I = \frac{Q}{t}$$

(3.9)

If a number of electrons with the total charge Q move with the velocity v in a conductor of length l, we have

$$Q = \frac{I \cdot l}{v}$$

(3.10)

---

[5] Wb, Wilhelm Eduard, German physicist, *1804, †1891.
[6]Hendrik Antoon Lorentz, Dutch mathematician and physicist,*1853, †1928.

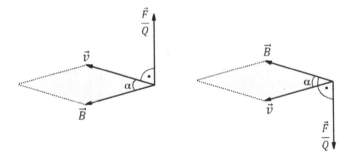

Fig. 3.6 Vector product

i.e.,

$$Q \cdot v = I \cdot l \tag{3.11}$$

Inserting (3.11) into (3.1)

$$\left|\vec{F}\right| = F = Q \cdot (v \cdot B \cdot \sin \alpha) \tag{3.12}$$

Equation (3.1) was derived for the homogeneous magnetic field inside an excitation inductor according to Fig. 3.4.

Equation (3.12) can be written in vector form. The magnetic flux density and the velocity are vectors, with the velocity vector pointing into the direction of the current. The force vector $\vec{F}$ is perpendicular to the surface formed by the velocity vector $\vec{v}$ and the vector of the magnetic flux density $\vec{B}$. In general, the flux vector $\vec{B}$ and the velocity vector $\vec{v}$ enclose an angle α. This relationship is mathematically expressed by the cross product (also called vector product) of the two vectors $\vec{v}$ and $\vec{B}$

$$\vec{F} = Q \cdot \left(\vec{v} \times \vec{B}\right) \tag{3.13}$$

In this equation, the vectors $\vec{v}$, $\vec{B}$, and $\vec{F}$ or $\vec{F}/Q$ form a right-hand screw (see Fn. 1) (Fig. 3.6).

According to (3.13), the length of the vector $\vec{F}/Q$ is equal to the surface of the parallelogram formed by the two vectors $\vec{v}$ and $\vec{B}$. At this point, it should be noted that the order of the two vectors $\vec{v}$ and $\vec{B}$ when the cross product is calculated must be observed.

3.1.1   Moving Conductor in a Stationary Magnetic Field

In the previous section, it was postulated that the force acting on a conductor, through which a current I flows, is generated by the force exerted by the magnetic field on moving charges. Moving charges do not only result from a current flow, but will also

Fig. 3.7 Induction effect in a rod moved in a homogeneous magnetic field (The direction of the vector $\vec{L}$ ($L =$ Length of moving conductor) has the same direction as the force $\vec{F}$ acting on the positive charge carriers in the conductor)

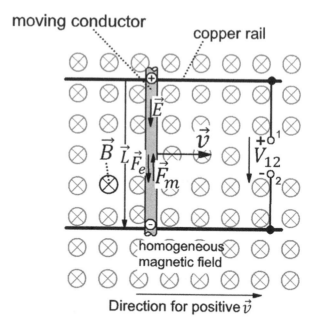

move if the conductor containing free electrons is being moved by an external force. Consequently, the free electrons in a moving conductor are displaced by the force of the magnetic field and thus generate a voltage inside the conductor.

Figure 3.7 shows a basic experimental setup demonstrating this effect. In this setup, a conductor of length $\left|\vec{L}\right| = L$ lies on two copper rails. It moves with the speed $\vec{v}$ in the indicated direction. The vector of the magnetic flux density $\vec{B}$ is oriented perpendicular to the direction of the motion of the conductor into the drawing plane[7]. The force $\vec{F}_m$ (corresponding to (3.13)) acts by the movement of the conductor on the positive charge carriers

$$\vec{F}_m = Q \cdot \left(\vec{v} \times \vec{B} \right)$$

The force $\vec{F}_m$ generated by the movement of the conductor causes an excess of positive charges at one end of the conductor (+ pole) and an excess of negative charges at the other end (− pole). An electric field $\vec{E}$ is built between the ends of the moving conductor, and therefore a voltage V_{12} can be measured at the terminals 1 and 2. It is called induced[8] voltage. If the electric circuit is closed, a current I will flow.

---

[7]In Fig. 3.7, a cross marks the end of the vector of arrow $\vec{B}$ that points into the drawing plane.
[8]Latin: inducere $=$ introduce.

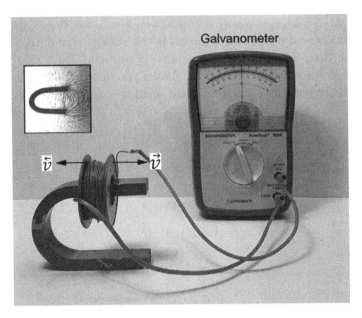

Fig. 3.8 Experimental arrangement for detecting the induction voltage and magnetic field of a horseshoe magnet

A magnetic field can be generated by a permanent magnet in the form of a horseshoe[9]. With an experimental arrangement as shown in Fig. 3.8, the induction voltage can thus be verified with a standard galvanometer. In order to generate a measurable voltage, an inductor with many windings is moved in the magnetic field instead of a single twist. As a result, there are many conductor sections in the area of the magnetic field, and the potential gradients of all these conductor sections add up. With such a simple test arrangement, voltage in the range of 20–50 mV can be achieved, depending on the speed of the movement and the number of windings.

According to (1.33), the voltage between the ends of the moving conductor in Fig. 3.7 is

$$V_{12} = \int_{1}^{2} \vec{E} \cdot \mathrm{d}\vec{l} \qquad (3.14)$$

---

[9]Permanent magnets are manufactured by pressing crystalline powder into shape under the influence of a strong magnetic field. The crystals align with their preferred magnetization axis in the direction of the magnetic field. Then, the presslings are sintered at a temperature of more than 1000° C, whereby the magnetic field is lost. After the magnets have cooled down, the magnetic field is restored by a sufficiently strong magnetization pulse. The magnetic field lines made visible by iron filings in Fig. 3.8 close within the magnetic material.

In this equation, $\vec{dl}$ is a length element of the moving conductor. The electric field counteracts the force exerted on the charge carriers by the movement of the conductor in the magnetic field. There is a balance of forces in the open circuit. The force F_m exerted by the movement in the magnetic field and the force F_e exerted by the electric field on the charge carriers are opposite to each other and equal in magnitude (see Fig. 3.7). From (3.13) and (1.31),

$$\vec{F}_m = Q \cdot \left(\vec{v} \times \vec{B} \right)$$
$$\vec{F}_e = Q \cdot \vec{E}$$
$$\vec{F}_m = -\vec{F}_e$$

$$Q \cdot \left(\vec{v} \times \vec{B} \right) = -Q \cdot \vec{E} \tag{3.15}$$

As a result

$$\vec{E} = -\left(\vec{v} \times \vec{B} \right) \tag{3.16}$$

Thus the voltage V_{12} is

$$V_{12} = \int\limits_1^2 \vec{E} \cdot \vec{dl} = -\int\limits_1^2 \left(\vec{v} \times \vec{B} \right) \cdot \vec{dl} = -\left(\vec{v} \times \vec{B} \right) \cdot \vec{L} \tag{3.17}$$

As the vectors $\vec{v}$ and $\vec{B}$ enclose an angle of 90° in the arrangement according to Fig. 3.7, (3.17) simplifies to

$$V_{12} = -|\vec{v}| \cdot \left| \vec{B} \right| \cdot \left| \vec{L} \right| = -v \cdot B \cdot L \tag{3.18}$$

3.1.2   Generator for Alternating Voltage

Figure 3.9 shows an arrangement with an additional second moving conductor. The two conductor rods moving on the copper rails form a conductor loop. They move at the speed $\vec{v}_1$, respectively, $\vec{v}_2$ in a homogeneous magnetic field. Between the ends of the two conductor rods, corresponding to (3.14), the voltages V_1 and V_2 will be generated.

The magnetic flux $d\Phi$ passing through a surface element $\vec{dA}$ located in the magnetic field with the flux density is

$$d\Phi = \vec{B} \cdot \vec{dA} \tag{3.19}$$

The vector $\vec{dA}$ is perpendicular to the surface element. The magnetic flux passing through a conductor loop with the surface A is referred to as the flux Φ_{con} concatenated to the conductor loop

$$\Phi_{\text{con}} = \iint\limits_A \vec{B} \cdot \vec{dA} \tag{3.20}$$

Fig. 3.9 Conductor loop in the static magnetic field

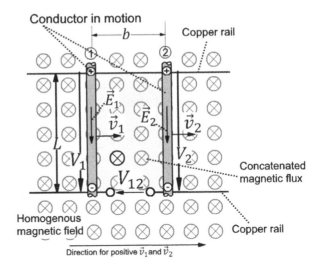

In Fig. 3.9, the surface A is formed by the two conductor rods and the copper rails connecting them. As the magnetic flux Φ_{con} is concatenated with this loop, we have

$$\Phi_{\text{con}} = \iint\limits_A \vec{B} \cdot d\vec{A} = (L \cdot b) \cdot B \tag{3.21}$$

With the designations and direction arrows defined in Figs. 3.7 and 3.9, we have according to (3.18)

$$V_2 = -v_2 \cdot B \cdot L = -\frac{ds_2}{dt} \cdot B \cdot L$$

$$V_1 = -v_1 \cdot B \cdot L = -\frac{ds_1}{dt} \cdot B \cdot L$$

The two conductor rods in Fig. 3.9 move at the speeds $v_1 = ds_1/dt$ and $v_2 = ds_2/dt$, respectively. If the velocity v_1 is smaller than v_2 for the same direction, the distance b between the two moving conductors in Fig. 3.9 will increase so that the magnetic flux $\Phi_{\text{con}}(t)$ concatenated to the loop 1–2–3–4 increases as well

$$\Phi_{\text{con}}(t) = L \cdot b(t) \cdot B$$

$$\frac{db(t)}{dt} = \frac{ds_2}{dt} - \frac{ds_1}{dt}$$

$$\frac{d\Phi_{\text{con}}}{dt} = L \cdot B \cdot \frac{db(t)}{dt} = L \cdot B \cdot \frac{ds_2}{dt} - L \cdot B \cdot \frac{ds_1}{dt}$$

$$\frac{d\Phi_{\text{con}}}{dt} = -V_2 + V_1$$

Now we have (see Fig. 3.9)

$$V_{12} = V_1 - V_2 = -\frac{d\Phi_{\text{con}}}{dt} \tag{3.22}$$

Fig. 3.10 Principle of a
generator for alternating
voltage

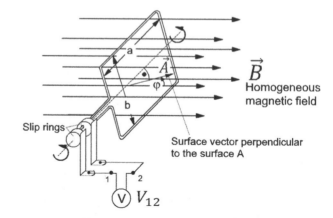

General statement of (3.22)

> A voltage will be generated at the terminals of a loop if the magnetic flux concate-
> nated with the loop changes with time. The cause of the time-varying concatenated
> magnetic flux is irrelevant.

A change of the flux concatenated with a loop can also be achieved by a loop rotating in
a magnetic field, as shown in Fig. 3.10. The axes of rotation and the flux density vector $\vec{B}$
are aligned perpendicular to each other. The loop rotates with the angular velocity ω. The
magnitude of the magnetic flux concatenated to the loop is thus dependent on the angle
of rotation φ. With $\varphi = 0$, the flux concatenated to the loop has its maximum value. With
$\varphi = 90°$, the flux concatenated to the loop is zero. The surface of the loop is $A = a \cdot b$.
The vector oriented perpendicular to this surface is the surface vector $\vec{A}$ with the magnitude A. For the projection of the surface A into the surface $A_\perp$ perpendicular to the direction of the magnetic flux density with $\varphi = \omega \cdot t$, we have

$$A_\perp = (a \cdot b) \cdot \cos \varphi = (a \cdot b) \cdot \cos(\omega \cdot t) \tag{3.23}$$

The magnetic flux Φ_{con} concatenated with the rotating loop therefore is time-dependent

$$\Phi_{con}(t) = \vec{B} \cdot \vec{A} = \left| \vec{B} \right| \cdot \left| \vec{A} \right| \cdot \cos(\omega \cdot t) \tag{3.24}$$

For voltage V_{12} present to terminals 1 and 2, with $\left(\left| \vec{B} \right| \cdot \left| \vec{A} \right| \cdot \omega = V_0 \right)$ according to
(3.22), we have

$$V_{12}(t) = -\frac{d\Phi_{con}}{dt} = \left| \vec{B} \right| \cdot \left| \vec{A} \right| \cdot \omega \cdot \sin(\omega \cdot t) = V_0 \cdot \sin(\omega \cdot t) \tag{3.25}$$

Fig. 3.11 Time function of
the voltage $V_{12}(t)$

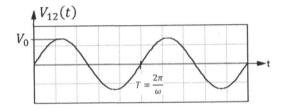

The time function of $V_{12}(t)$ is shown in Fig. 3.11. The arrangement in Fig. 3.10 thus represents the principle of a generator for alternating voltage.

3.1.3   Magnetic Flux Measurement

The magnetic flux can be measured with a Hall[10] sensor (see Fig. 3.12). A Hall sensor consists of a planar conductor with the thickness d, the width b, and the length l. A current I is fed into the front surface of the Hall sensor. When the Hall sensor is placed in a magnetic field with the flux density $\vec{B}$, oriented perpendicular to the planar conductor of the Hall sensor, a voltage will be generated between opposite points perpendicular to the current and the field direction. This voltage is called Hall voltage.

The current I flows through the entire volume of the Hall sensor. The moving electrons of this current are influenced by the Lorentz force deflecting them perpendicular to the direction of the current and the magnetic field. This results in an increased density of the electrons on one side of the Hall sensor and consequently results in an induced electric field E_i between the side surfaces of the sensor. The result is a transverse voltage, the Hall voltage V_H

Fig. 3.12 Hall sensor for
measuring magnetic fields

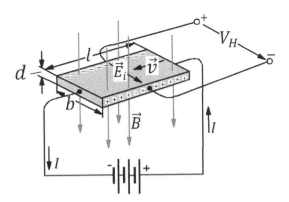

---

[10]Hall, Edwin H., American physicist, *1855, †1938.

$$V_H = -V_i = -E_i \cdot b \tag{3.26}$$

After (3.16), the induced electric field is

$$\vec{E}_i = -\left(\vec{v} \times \vec{B}\right) \tag{3.27}$$

Since the velocity vector is perpendicular to the direction of the vector of the magnetic flux density

$$E_i = -v \cdot B \tag{3.28}$$

With (3.26), we have

$$V_H = -V_i = -v \cdot B \cdot b \tag{3.29}$$

with the speed

$$v = \frac{s}{t}$$

In this equation, t is the time the charge carriers travel the distance s within the Hall sensor. The current I in Fig. 3.12 is

$$I = \frac{Q}{t} = \frac{Q}{s} \cdot v = \frac{Q \cdot b \cdot d}{s \cdot b \cdot d} \cdot v \tag{3.30}$$

Q is the quantity of charge which moves within the Hall sensor. The volume of the Hall sensor is $(s \cdot b \cdot d)$. Thus, the space charge density is

$$\varrho = \frac{Q}{s \cdot b \cdot d} \tag{3.31}$$

Consequently, the current I is

$$I = \varrho \cdot v \cdot b \cdot d \tag{3.32}$$

Therefore

$$v = \frac{I}{\varrho \cdot b \cdot d} \tag{3.33}$$

With (3.29), the Hall voltage V_H is

$$V_H = -V_i = -v \cdot B \cdot b = -\frac{I}{\varrho \cdot b \cdot d} \cdot B \cdot b$$

or

$$V_H = -\frac{1}{\varrho} \cdot \frac{I \cdot B}{d} \tag{3.34}$$

Table 3.1 Hall constants of some materials

Material	Hall constant R_H in $\frac{cm^3}{A \cdot s}$
Metals	10^{-4}
Germanium	10^3
Silicon	10^6
Indium arsenide	$14-30$
Gallium arsenide	$50-85$

The factor $1/\varrho$ in (3.34) is called the Hall constant R_H, thus

$$V_H = -R_H \cdot \frac{I \cdot B}{d} \tag{3.35}$$

From the measured values of current I and Hall voltage V_H, the magnetic flux density B can be determined

$$|B| = \frac{V_H}{R_H \cdot I} \cdot d \tag{3.36}$$

At constant current I, the Hall voltage V_H increases with reducing layer thickness d of the Hall probe. The typical layer thickness is between $1\,\mu m$ and some $10\,\mu m$. The Hall constant is quite small for metals, but particularly high for semiconductors (see Table 3.1). Pure semiconductors such as germanium and silicon have a very large Hall constant, but also a very large specific resistance. Semiconductor alloys such as indium arsenide and gallium arsenide are better suited for technical applications, because of a small Hall constant and a small specific resistance.

3.2   Ampère's Law[11]

The subject of this section is the relationship between the magnetic field and the electric current that generates this magnetic field. First, a measuring arrangement is presented verifying this relationship, which is referred to as Ampère's law.

A so-called Rogowski inductor[12] or Rogowski coil is used for the measurement. It is an elongated, flexible inductor with a small cross section that is wound as uniformly as possible around a flexible inductor core made out of non-magnetic material (ironless inductor). Figure 3.13 on the right shows the basic structure of the inductor. In this figure, the inductor is bent into a circle, open at the ends a and b. The left part of the figure shows the view of a non-bent Rogowski inductor. We can see that their windings are very close to each other.

---

[11]André-Marie Ampère, French physicist and mathematician, *1775, †1836.

[12]Rogowski, W., German electrical engineer, *1881, †1947.

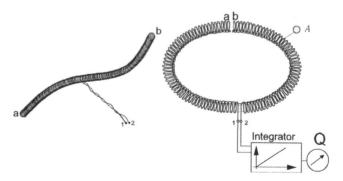

Fig. 3.13 Rogowski inductor

Fig. 3.14 Schematic diagram of an ideal integration amplifier

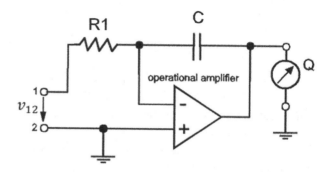

For the measurement, the connections 1 and 2 of the inductor are connected to an integrating amplifier. Its principle circuit diagram is shown in Fig. 3.14. The measured value Q is proportional to the integral of the time-dependent voltage $v_{12}(t)$ at the terminals 1 and 2 of the Rogowski inductor.

$$Q = K \cdot \int v_{12}(t) \cdot dt \tag{3.37}$$

$K =$ Proportionality factor

The cross section of the Rogowski inductor is small, so that the magnetic flux density in the cross section can be considered as constant. The cross-sectional surface of the Rogowski inductor is A, and n is the number of windings per unit length. Thus, the relationship between the specific magnetic flux concatenated with the inductor per length element $d\vec{s}$ and the flux density $\vec{B}$ is (cf. (3.19))

$$d\Phi_{\text{con/spec}} = n \cdot A \cdot \vec{B} \cdot d\vec{s} \tag{3.38}$$

The vector $d\vec{s}$ has the direction of the tangent to the curved inductor axis of the Rogowski inductor. The magnetic flux, concatenated in total with the Rogowski inductor, therefore is

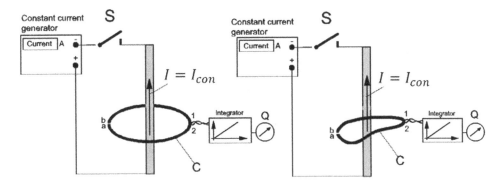

Fig. 3.15 Measurements with the Rogowski inductor

$$\Phi_{\text{con}} = n \cdot A \cdot \int_1^2 \vec{B} \cdot d\vec{s} = n \cdot A \cdot \oint_C \vec{B} \cdot d\vec{s} \tag{3.39}$$

C is the contour of the Rogowski inductor.

Figure 3.15 shows the measurement setup for the verification of Ampère's law. The Rogowski inductor encloses the current-carrying conductor (The two ends a and b of the inductor are close together).

If the switch S is closed, a current I will flow through the conductor. This current increases with time to its final value. This creates a time-varying magnetic field around the conductor. The magnetic flux concatenated with the Rogowski inductor increases from zero, before closing the switch, to the final value Φ_{con}. As a result, according to (3.22), a voltage is induced in the Rogowski inductor[13]. With (3.22), the voltage, induced at terminals 1 and 2 of the Rogowski inductor at the time t, is[14]

$$\frac{d\Phi_{\text{con}}}{dt} = v_{12}(t) \tag{3.40}$$

With (3.39), we get

$$\frac{d\Phi_{\text{con}}}{dt} = v_{12}(t) = \frac{d}{dt}\left(n \cdot A \cdot \oint_C \vec{B} \cdot d\vec{s}\right) \tag{3.41}$$

---

[13]The cause of the time-varying, concatenated magnetic flux $d\Phi_{\text{con}}/dt$ in Eq. (3.40) is the current, which increases with time during the switch-on process. The time-varying, concatenated magnetic flux in Eq. (3.22) is caused by the changing cross section of the conductor loop. In order for the induced voltage to arise, it is indifferent whereby the time-varying, concatenated flux arises. In detail, time-varying magnetic fields are dealt with in Sect. 4.1.

[14]The sign of the voltage v_{12} is irrelevant in this case.

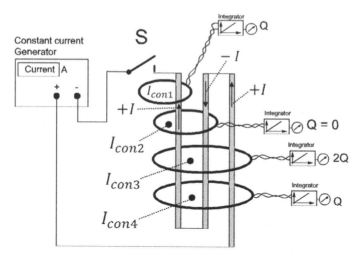

Fig. 3.16 Rogowski inductor with four different values of current, concatenated with the inductor

The integral in (3.41) extends over the closed contour C of the Rogowski inductor enclosing the current-carrying conductor between the ends a and b. Since the rate of change of the concatenated magnetic flux is not constant, also the voltage $v_{12}(t)$ is not constant with time. If the current flow has reached its final value I, the voltage $v_{12}(t)$ will drop to zero. The displayed value Q at the output of the integrator is according to (3.37) proportional to the integral of the voltage $v_{12}(t)$. With (3.41) in conjunction with (3.37), we have

$$Q = K \cdot \int_T v_{12}(t) \cdot \mathrm{d}t = K \cdot \frac{n \cdot A}{R} \cdot \oint_C \vec{B} \cdot \mathrm{d}\vec{s} \tag{3.42}$$

The shape of the closed contour C of the Rogowski loop does not matter. The measurement results with the Rogowski inductor in the left part of Fig. 3.15 are identical to the results in the right part of Fig. 3.15.

Figure 3.16 shows four different values I_{con1} to I_{con4} of current concatenated with a Rogowski inductor. The results of the measurements are summarized in Table 3.2.

Table 3.2 Measurement results of Fig. 3.16

Measurement	Concatenated current	Display of the integrator
1.	$I_{con1} = I$	Q_1
2.	$I_{con2} = I - I = 0$	0
3.	$I_{con3} = I + I = 2 \cdot I$	$2 \cdot Q_1$
4.	$I_{con4} = I - I + I = I$	Q_1

General result:

> The integral of the magnetic flux density along a closed contour C enclosing the current-carrying conductor, i.e., the magnetic flux Φ_{con} concatenated with this contour, is proportional to the current $I = I_{\text{con}}$ passing through the loop C.

In (3.43), this fact is formulated mathematically

$$\oint_C \vec{B} \cdot d\vec{s} = \Phi_{\text{con}} = \mu \cdot I_{\text{con}} = \mu \cdot I \tag{3.43}$$

In (3.43), μ is the proportionality factor. It is called permeability. Experiments have shown that permeability depends on the medium, in which the magnetic field exists. In Fig. 3.16, the medium is air. The permeability μ is composed of two factors:

$$\mu = \mu_r \cdot \mu_0 \tag{3.44}$$

In (3.44), μ_0 is the permeability of vacuum, which is nearly identic to the permeability of air. It is called absolute permeability. In contrast to this, the relative permeability μ_r describes the magnetic properties of the material in which the magnetic field exists.

The value of the absolute permeability μ_0 is determined by the definition of the electric current. Reformulating (3.43), we get the following relationship, which is called Ampère's law:

$$\oint_C \frac{\vec{B}}{\mu} \cdot d\vec{s} = \oint_C \vec{H} \cdot d\vec{s} = I \tag{3.45}$$

In this equation, the vector

$$\vec{H} = \frac{\vec{B}}{\mu} \tag{3.46}$$

is called magnetic field.

Ampère's law in text form:

> In any magnetic field, the integral of the magnetic field $\vec{H}$ along a closed contour C is equal to the total current I passing through the surface enclosed by this contour.

3.2.1 Absolute Permeability

To illustrate Ampère's law, Fig. 3.17 shows the cross section of an infinitely long conductor, through which a current I flows and one of the circular magnetic field lines surrounding this conductor. In the example, we will integrate along the circular field line.

Fig. 3.17 Ampère's law, illustrated on an infinitely long, straight, current-carrying conductor

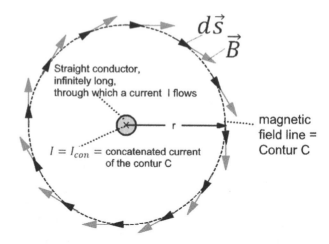

The value of absolute permeability μ_0 is directly related to the definition of the electric current. The electric current unit Ampere (symbol A) is an SI base unit and is defined as follows:

1 Ampere (A) is the intensity of a steady electric current producing a specific force of $2 \cdot 10^{-7}$ N/m in a vacuum between two parallel, infinitely long, straight conductors with a negligibly small circular cross section and a distance of 1 m between these conductors.[15]

The definition of the unit Ampere of the electric current was traditionally linked to the definition of the unit of the magnetic flux density B:

The magnetic flux density is 1 T (Tesla), when a measuring rod of length 1 m, through which a current of A flows (see (3.3)), exerts a force of 1 N.

---

[15]From 20 May 2019, a new definition of the Ampere unit in the International System of Units (SI) applies. The new definition of the unit Ampere is based on the precisely defined value of the elementary charge e. It was possible to change the definition, as individual charge can now be counted well. According to the new definition, the unit 1 A is present if $1,602176634 \cdot 10^9$ elementary charges flow through the conductor during 1 s. As a result of this definition, the field constants μ_0, ε_0 and the characteristic impedance of vacuum are now derived variables subject to uncertainty (Source: Wikipedia).

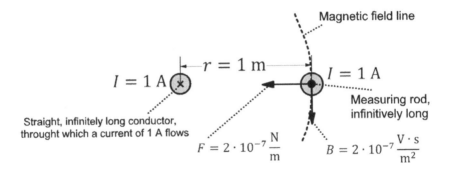

Fig. 3.18 On the value of absolute permeability μ_0

Figure 3.18 shows this relationship in a drawing.

Due to the definition of intensity of the electrical current, a specific force of

$$F = 2 \cdot 10^{-7} \frac{N}{m}$$

is exercised on the measuring rod in Fig. 3.18.

In accordance with (3.3), the magnetic flux density at the point of the measuring rod by definition is

$$B = 2 \cdot 10^{-7} \frac{N}{A \cdot m} = 2 \cdot 10^{-7} \cdot \frac{V \cdot s}{m^2}$$

The current in the center of the circle with a radius of 1 m is 1 A. Thus, after (3.43)

$$\oint_C \vec{B} \cdot d\vec{s} = 2 \cdot 10^{-7} \cdot \frac{V \cdot s}{m^2} \cdot 2 \cdot \pi \cdot 1 \cdot m = \mu_0 \cdot 1 \cdot A$$

Consequently, the absolute permeability is

$$\mu_0 = 2 \cdot 10^{-7} \cdot \frac{\frac{V \cdot s}{m^2} \cdot 2 \cdot \pi \cdot 1 \cdot m}{1 \cdot A} = 4 \cdot \pi \cdot 10^{-7} \cdot \frac{V \cdot s}{A \cdot m} \tag{3.47}$$

3.2.2 Magnetic Field Inside and Outside of an Infinitely Long Conductor

As an example, the magnetic field $\vec{H}$ is calculated inside and outside of a conductor, which is infinitely long (see Fig. 3.19).

Fig. 3.19 Section of a
straight, infinitely long
conductor with current

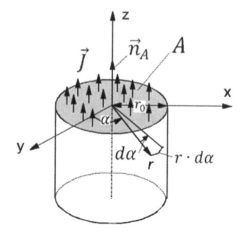

Current within the conductor:	I
Current density within the conductor:	$\vec{J}$
Radius of the conductor:	r_0

The current density $\vec{J}$ is constant over the cross section of the conductor.

Solution:

The magnetic field lines are circles around the axis of the conductor inside and outside of the current-carrying conductor. As a result, the magnetic field has only a component in the α-direction and does not depend on the angle α. For surfaces within the conductor, i.e., for surfaces in which a current density $\vec{J}$ is present, the Ampère's law says (see (3.45))

$$\oint_C \vec{H} \cdot d\vec{s} = \iint_A \vec{J} \cdot \vec{n}_A \cdot dA \tag{3.48}$$

$$\oint_{\alpha=0}^{2\cdot\pi} H_{\alpha/in}(r) \cdot r \cdot d\alpha = \int_{\alpha=0}^{2\cdot\pi} \int_{r=0}^{r} \frac{I}{\pi \cdot r_0^2} \cdot r \cdot d\alpha \cdot dr$$

$$H_{\alpha/in}(r) \cdot 2 \cdot \pi \cdot r = \frac{I}{\pi \cdot r_0^2} \cdot 2 \cdot \pi \cdot \int_0^r r \cdot dr$$

$$H_{\alpha/in}(r) \cdot r = \frac{I}{\pi \cdot r_0^2} \cdot \frac{r^2}{2}$$

$$H_{\alpha/in}(r) = \frac{I}{2 \cdot \pi} \cdot \frac{r}{r_0^2} \tag{3.49}$$

Surface outside of the conductor

$$\oint_{\alpha=0}^{2\cdot\pi} H_{\alpha/\mathrm{out}}(r) \cdot r \cdot \mathrm{d}\alpha = \iint_A \vec{J} \cdot \vec{n}_A \cdot \mathrm{d}A = I$$

$$H_{\alpha/\mathrm{out}}(r) \cdot 2 \cdot \pi \cdot r = I$$

$$H_{\alpha/\mathrm{out}}(r) = \frac{I}{2 \cdot \pi} \cdot \frac{1}{r} \qquad (3.50)$$

Within the conductor, the magnetic field increases linearly with the distance r from the center of the conductor. Outside of the conductor, the magnetic field is proportional to $1/r$.

3.3   Magnetic Scalar Potential

If the contour does not include any current according to Ampère's law, the integral of the magnetic field over a closed contour is zero. In this case (see (3.45)),

$$\oint_C \frac{\vec{B}}{\mu} \cdot \mathrm{d}\vec{s} = \oint_C \vec{H} \cdot \mathrm{d}\vec{s} = 0 \qquad (3.51)$$

If the integral of the magnetic field is performed between two points a and b of a magnetic field, then the value of the integral is independent of the integration path

$$\int_a^b \vec{H} \cdot \mathrm{d}\vec{s} = \text{independent of the integration path}$$

Two integration paths between the points a and b are shown in Fig. 3.20. The integration paths do not enclose or cut any current-carrying conductor. To prove the independence of the integral from the integration path, two paths with their line integral are compared in Fig. 3.20.

Integral via path 1 (index 1)

$$\int_{a_1}^{b_1} \vec{H} \cdot \mathrm{d}\vec{s}$$

Integral via path 2 (index 2)

$$\int_{a_2}^{b_2} \vec{H} \cdot \mathrm{d}\vec{s}$$

Fig. 3.20 Integration paths of the integral of the magnetic field

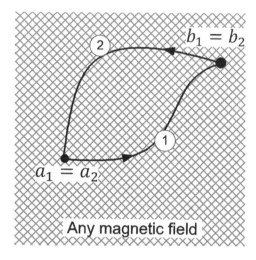

The sum of the integral from a_1 to b_1 via path 1 and of the integral back via path 2 from b_2 to a_2 is equal to zero if, as assumed, the contour formed by the paths 1 and 2 does not include a current-carrying conductor

$$\int_{a_1}^{b_1} \vec{H} \cdot d\vec{s} + \int_{b_2}^{a_2} \vec{H} \cdot d\vec{s} = 0$$

It follows

$$\int_{a_1}^{b_1} \vec{H} \cdot d\vec{s} = \int_{a_2}^{b_2} \vec{H} \cdot d\vec{s}$$

In text form:

Provided that the paths between two points can be merged without cutting a current conducting conductor, the integral of the magnetic field between two points a and b is independent of the integration path between these points. The integral depends only on the position of the two points a and b in the magnetic field (quoted after [4]).

This fact has its analogy in the electric field (cf. (1.15)). In analogy to the electric scalar potential, a magnetic scalar potential ψ can be introduced defined by the following relationship (cf. (1.16) and (1.6)):

$$\int_{a}^{b} \vec{H} \cdot d\vec{s} = \psi_a - \psi_b \tag{3.52}$$

or

$$\vec{H} = -\text{grad}\,\psi = -\nabla\psi \tag{3.53}$$

This means that outside of current-carrying conductors, a potential value can be assigned to each point of the magnetic field. The magnet field is equal to the gradient of this scalar field.

The integral of the magnetic field therefore is called magnetic voltage in analogy to the electrical voltage. The magnetic voltage can be measured with the Rogowski inductor according to Fig. 3.13 (see also (3.39)). For this reason, the Rogowski inductor is also referred to as a magnetic voltmeter.

3.4 Differential Form of Ampère's Law

The integral form of Ampère's law according to (3.45) specifies the relationship between the magnetic field along a contour and the current passing through the surface enclosed by this contour. In order to find the relationship between the magnetic field at a point of the field within a current-carrying conductor and the current density present there, Ampère's law must be converted from its integral form into the differential form.

The starting point is (3.48). The integration extends over the surface A enclosed by the contour C (The vector $\vec{J}$ and the direction of the rotation in the direction $\mathrm{d}\vec{s}$ form a right-hand screw) (Fig. 3.21)

$$\oint_C \vec{H} \cdot \mathrm{d}\vec{s} = \iint_A \vec{J} \cdot \vec{n}_A \cdot \mathrm{d}A \tag{3.54}$$

In (Fig. 3.21), a surface A with the contour C is depicted. Within the contour the current density $\vec{J}$ is location dependent. If in (3.54) the surface A is reduced to a very small surface ΔA, the current density $\vec{J}$ can be assumed to be constant in this surface.

With the small surface ΔA (3.54) changes into

$$\oint_{C_{\Delta A}} \vec{H} \cdot \mathrm{d}\vec{s} = \vec{J} \cdot \Delta A \cdot \vec{n}_{\Delta A} \tag{3.55}$$

Fig. 3.21 On the differential form of Ampère's law

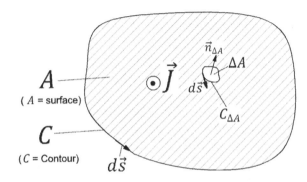

In (3.55), $C_{\Delta A}$ is the contour of the small surface ΔA. Dividing both sides of this equation by ΔA yields

$$\frac{1}{\Delta A} \cdot \oint_{C_{\Delta A}} \vec{H} \cdot d\vec{s} = \vec{J} \cdot \vec{n}_{\Delta A} \tag{3.56}$$

Multiplying both sides with the unit vector $\vec{n}_{\Delta A}$ gives a relationship with only the current density $\vec{J}$ on the right side

$$\frac{\vec{n}_{\Delta A}}{\Delta A} \cdot \oint_{C_{\Delta A}} \vec{H} \cdot d\vec{s} = \vec{J} \tag{3.57}$$

To explicitly express that the small surface ΔA is infinitesimal, we use the limit

$$\vec{n}_{\Delta A} \cdot \lim_{\Delta A \to 0} \frac{1}{\Delta A} \cdot \oint_{C_{\Delta A}} \vec{H} \cdot d\vec{s} = \vec{J} \tag{3.58}$$

In vector analysis, the limit of the integral of a vector along a closed infinitesimal contour $C_{\Delta A}$ divided by the infinitesimally small surface ΔA formed by the contour is called the rotation or curl of that vector. This vector operation, applied to the vector field $\vec{H}$, has the name curl $\vec{H}$. Thus with (3.58)

$$\operatorname{curl} \vec{H} = \vec{J} \tag{3.59}$$

Equation (3.59) is the differential form of Ampère's law. The designation curl $\vec{H} \neq 0$ expresses the fact that the magnetic field line at the respective location, to which this vector function is applied, has a vortex (curl). Instead of the spelling "curl", a mathematical equivalent expression with the Nabla operator is more common. It is called "del cross"

$$\operatorname{curl} \vec{H} = \nabla \times \vec{H} = \vec{J} \tag{3.60}$$

3.5   Operation $\nabla \times$ Applied to the Magnetic Field $\vec{H}$

3.5.1   Vector Operation Curl $\vec{H} = \nabla \times \vec{H}$ in Cylindrical Coordinates

To explain the vector operation ($\nabla \times$) on the vector field $\vec{H}$, the curl of the magnetic field of an infinitely long, current-carrying conductor, as shown in Fig. 3.19, is calculated as an example. This example is a cylinder-symmetric problem. For this reason, the components of the vector operation curl $\vec{H} = \nabla \times \vec{H}$ are calculated in cylinder coordinates.

To determine the components curl$_r$ $\vec{H}$, curl$_\alpha$ $\vec{H}$, and curl$_z$ $\vec{H}$ of the vector $\nabla \times \vec{H}$, the operation described above is to be performed along the three components $\mathrm{d}A_r$, $\mathrm{d}A_\alpha$, and

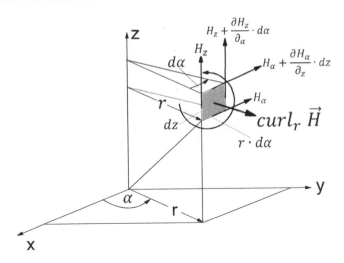

Fig. 3.22 r-component curl$_r$ $\vec{H}$ of the vector $\nabla \times \vec{H}$ and component dA_r of the surface element d$\vec{A}$ for $r=$ const

dA_z of the surface d$\vec{A}$. Then, the result has to be divided by the corresponding component of dA.

The component dA_r is the surface element, resulting from d$\vec{A}$ for $r =$ const if the angle α is changed by dα and the coordinate z is changed by dz. This surface element is shown in Fig. 3.22. The component curl$_r$ $\vec{H}$ is perpendicular to this surface element and points into the positive r-direction. For the calculation of the component curl$_r$ $\vec{H}$, the component dA_r must be circumnavigated in the sense of a right-hand screw with respect to curl$_r$ $\vec{H}$.

The variation of the z-component of the magnetic field during progression in α-direction is

$$\frac{\partial H_z}{\partial \alpha} \cdot d\alpha$$

Accordingly, we have the variation of the α-component during progression in the z-direction as

$$\frac{\partial H_\alpha}{\partial z} \cdot dz$$

Figure 3.22 shows the following relationship for the integral $\left(\oint \vec{H} \cdot d\vec{s}\right)_{r=\text{const}}$:

$$\left(\oint \vec{H} \cdot d\vec{s}\right)_{r=\text{const}}$$
$$= H_\alpha \cdot r \cdot d\alpha + \left(H_z + \frac{\partial H_z}{\partial \alpha} \cdot d\alpha\right) \cdot dz - \left(H_\alpha + \frac{\partial H_\alpha}{\partial z} \cdot dz\right) \cdot r \cdot d\alpha - H_z \cdot dz \tag{3.61}$$

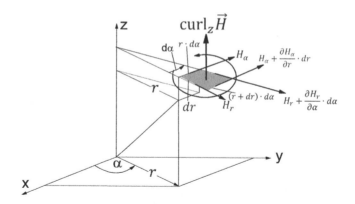

Fig. 3.23 z-component curl$_z\,\vec{H}$ of the vector curl $\vec{H} = \nabla \times \vec{H}$ and component dA$_z$ of the surface element dA for $z = $ const

$$\left(\oint \vec{H} \cdot d\vec{s}\right)_{r=\text{const}}$$

$$= H_\alpha \cdot r \cdot d\alpha + H_z \cdot dz + \frac{\partial H_z}{\partial \alpha} \cdot d\alpha \cdot dz - H_\alpha \cdot r \cdot d\alpha - \frac{\partial H_\alpha}{\partial z} \cdot dz \cdot r \cdot d\alpha - H_z \cdot dz$$

$$\left(\oint \vec{H} \cdot d\vec{s}\right)_{r=\text{const}} = \left(\frac{\partial H_z}{\partial \alpha} - \frac{r \cdot \partial H_\alpha}{\partial z}\right) \cdot d\alpha \cdot dz$$

Since the r-component dA$_r$ of the surface element d$\vec{A}$ is equal to $(r \cdot d\alpha \cdot dz)$, we obtain the r-component of the vector operation curl $\vec{H} = \nabla \times \vec{H}$ and (3.59)

$$\text{curl}_r\,\vec{H} = \frac{1}{r \cdot d\alpha \cdot dz} \cdot \left(\oint \vec{H} \cdot d\vec{s}\right)_{r=\text{const}} = \frac{1}{r} \cdot \frac{\partial H_z}{\partial \alpha} - \frac{\partial H_\alpha}{\partial z} \qquad (3.62)$$

The calculation of the component curl$_z\,\vec{H}$ is carried out according to the same scheme. The following equation can be derived from Fig. 3.23:

$$\left(\oint \vec{H} \cdot d\vec{s}\right)_{z=\text{const}}$$

$$= H_r \cdot dr + \left(H_\alpha + \frac{\partial H_a}{\partial r} \cdot dr\right) \cdot (r + dr) \cdot d\alpha - \left(H_r + \frac{\partial H_r}{\partial \alpha} \cdot d\alpha\right) \cdot dr - H_\alpha \cdot r \cdot d\alpha$$

$$(3.63)$$

$$\left(\oint \vec{H} \cdot d\vec{s}\right)_{z=\text{const}}$$

$$= H_r \cdot dr + H_\alpha \cdot r \cdot d\alpha + H_\alpha \cdot dr \cdot d\alpha + \frac{\partial H_\alpha}{\partial r} \cdot dr \cdot r \cdot d\alpha$$

$$+ \frac{\partial H_\alpha}{\partial r} \cdot dr \cdot dr \cdot d\alpha - H_r \cdot dr - \frac{\partial H_r}{\partial \alpha} \cdot d\alpha \cdot dr - H_\alpha \cdot r \cdot d\alpha$$

$$\left(\oint \vec{H} \cdot d\vec{s}\right)_{z=\text{const}}$$

$$= H_\alpha \cdot dr \cdot d\alpha + \frac{\partial H_\alpha}{\partial r} \cdot dr \cdot r \cdot d\alpha + \frac{\partial H_\alpha}{\partial r} \cdot dr \cdot dr \cdot d\alpha - \frac{\partial H_r}{\partial \alpha} \cdot d\alpha \cdot dr$$

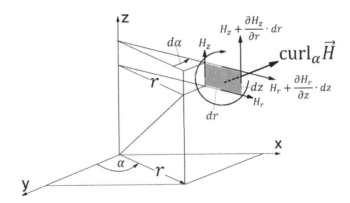

Fig. 3.24 α-component $\mathrm{curl}_\alpha \vec{H}$ of the vector $\mathrm{curl}\,\vec{H} = \nabla \times \vec{H}$ and component $\mathrm{d}A_\alpha$ of the surface element $\alpha = \mathrm{const}$

The expression $\frac{\partial H_\alpha}{\partial r} \cdot \mathrm{d}r \cdot \mathrm{d}r \cdot \mathrm{d}\alpha$ can be neglected.

$$\left(\oint \vec{H} \cdot \mathrm{d}\vec{s}\right)_{z=\mathrm{const}} = H_\alpha \cdot \mathrm{d}r \cdot \mathrm{d}\alpha + \frac{\partial H_\alpha}{\partial r} \cdot \mathrm{d}r \cdot r \cdot \mathrm{d}\alpha - \frac{\partial H_r}{\partial \alpha} \cdot \mathrm{d}\alpha \cdot \mathrm{d}r$$

$$\left(\oint \vec{H} \cdot \mathrm{d}\vec{s}\right)_{z=\mathrm{const}} = \left(\frac{1}{r}H_\alpha + \frac{\partial H_\alpha}{\partial r} \cdot - \frac{1}{r}\frac{\partial H_r}{\partial \alpha}\right) \cdot (\mathrm{d}r \cdot r \cdot \mathrm{d}a)$$

$$\left(\oint \vec{H} \cdot \mathrm{d}\vec{s}\right)_{z=\mathrm{const}} = \left[\frac{1}{r} \cdot \left(\frac{\partial (r \cdot H_\alpha)}{\partial r} - \frac{\partial H_r}{\partial \alpha}\right)\right] \cdot (\mathrm{d}r \cdot r \cdot \mathrm{d}a)$$

Since the z-component $\mathrm{d}A_z$ of the surface element $\mathrm{d}\vec{A}$ is equal to $(r \cdot \mathrm{d}\alpha \cdot \mathrm{d}r)$, we get corresponding to

(3.58) and (3.59) the z-component $\mathrm{curl}_z \vec{H}$ of the vector operation $\mathrm{curl}\,\vec{H} = \nabla \times \vec{H}$

$$\mathrm{curl}_z \vec{H} = \frac{1}{\mathrm{d}r \cdot r \cdot \mathrm{d}\alpha} \cdot \left(\oint \vec{H} \cdot \mathrm{d}\vec{s}\right)_{z=\mathrm{const}} = \frac{1}{r}\left(\frac{\partial (r \cdot H_\alpha)}{\partial r} - \frac{\partial H_r}{\partial \alpha}\right) \qquad (3.64)$$

The calculation of the component $\mathrm{curl}_\alpha \vec{H}$ is performed on the basis of Fig. 3.24. The following relationship is obtained:

$$\left(\oint \vec{H} \cdot \mathrm{d}\vec{s}\right)_{\alpha=\mathrm{const}}$$
$$= H_z \cdot \mathrm{d}z + \left(H_r + \frac{\partial H_r}{\partial z} \cdot \mathrm{d}z\right) \cdot \mathrm{d}r - \left(H_z + \frac{\partial H_z}{\partial r} \cdot \mathrm{d}r\right) \cdot \mathrm{d}z - H_r \cdot \mathrm{d}r \qquad (3.65)$$

$$\left(\oint \vec{H} \cdot \mathrm{d}\vec{s}\right)_{\alpha=\mathrm{const}} = H_z \cdot \mathrm{d}z + H_r \cdot \mathrm{d}r + \frac{\partial H_r}{\partial z} \cdot \mathrm{d}r \cdot \mathrm{d}z - H_z \cdot \mathrm{d}z - \frac{\partial H_z}{\partial r} \cdot \mathrm{d}r \cdot \mathrm{d}z - H_r \cdot \mathrm{d}r$$

$$\left(\oint \vec{H} \cdot \mathrm{d}\vec{s}\right)_{\alpha=\mathrm{const}} = \left(\frac{\partial H_r}{\partial z} - \frac{\partial H_z}{\partial r}\right) \cdot \mathrm{d}r \cdot \mathrm{d}z$$

Since the α-component $\mathrm{d}A_\alpha$ of the surface element $\mathrm{d}\vec{A}$ is equal to $(\mathrm{d}z \cdot \mathrm{d}r)$, the α-component of the vector operation $\mathrm{curl}\,\vec{H} = \nabla \times \vec{H}$ is obtained according to (3.58) and (3.59) as

$$\mathrm{curl}_\alpha\,\vec{H} = \frac{1}{\mathrm{d}r \cdot \mathrm{d}z} \cdot \left(\oint \vec{H} \cdot \mathrm{d}\vec{s}\right)_{\alpha=\mathrm{const}} = \frac{\partial H_r}{\partial z} - \frac{\partial H_z}{\partial r} \tag{3.66}$$

Equations (3.62), (3.64), and (3.66) can be summarized using the Nabla operator ∇ in the following matrix notation:

$$\mathrm{curl}\,\vec{H} = \nabla \times \vec{H} = \frac{1}{r}\begin{vmatrix} \vec{e}_r & r \cdot \vec{e}_\alpha & \vec{e}_z \\ \frac{\partial}{\partial r} & \frac{\partial}{\partial \alpha} & \frac{\partial}{\partial z} \\ H_r & r \cdot H_\alpha & H_z \end{vmatrix} \tag{3.67}$$

According to the rules for three-row determinants, we have [7] ($\vec{e}_x, \vec{e}_\alpha, \vec{e}_z$ are the unit vectors).

$$\nabla \times \vec{H} = \frac{1}{r}\vec{e}_r\left(\frac{\partial \vec{H}_z}{\partial \alpha} - \frac{\partial\left(r \cdot \vec{H}_\alpha\right)}{\partial z}\right) - \frac{1}{r}(r \cdot \vec{e}_\alpha)\left(\frac{\partial \vec{H}_z}{\partial r} - \frac{\partial \vec{H}_r}{\partial z}\right) + \frac{1}{r}\vec{e}_z\left(\frac{\partial\left(r \cdot \vec{H}_\alpha\right)}{\partial r} - \frac{\partial \vec{H}_r}{\partial \alpha}\right)$$

i.e.,

$$\nabla \times \vec{H} = \left(\frac{1}{r} \cdot \frac{\partial \vec{H}_z}{\partial \alpha} - \frac{\partial \vec{H}_\alpha}{\partial z}\right)\vec{e}_r - \left(\frac{\partial \vec{H}_z}{\partial r} - \frac{\partial \vec{H}_r}{\partial z}\right)\vec{e}_\alpha + \left(\frac{\partial \vec{H}_\alpha}{\partial r} - \frac{1}{r}\frac{\partial \vec{H}_r}{\partial \alpha}\right)\vec{e}_z \tag{3.68}$$

3.5.2   Circulation of the Vector Field of an Infinitely Long Conductor

Using the example of the magnetic vector field, calculated in Sect. 3.2.2, the essence of the vector operation $\left(\nabla \times \vec{H}\right)$ is examined in more detail below.

To form the curl operation of the magnetic vector field, the scalar product of the path element and the magnetic field vector must be calculated at all points of the field, when circumnavigating the surface element. If the result is positive, the direction of the magnetic field is on average aligned in the same way as the circumnavigation of the surface element. Then the field has in this case vortex, respectively, a curl, at this location and according to Ampère's law, the surface element at this location, is permeated by a current density, whose value corresponds to the curl of the vector field.

According to (3.49) and (3.50), the magnetic field of the infinitely long current-carrying conductor of Fig. 3.19 has only a component in the α-direction, depending only on the r-coordinate.

For the calculation of $\nabla \times \vec{H}$, therefore only (3.64) and only the part containing the partial differentiation of the component H_α with respect to r, that is, $\partial H_\alpha/\partial r$ is to be calculated

Fig. 3.25 Surface element of
the conductor cross section

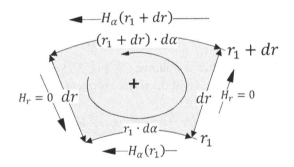

$$\text{curl}_z \, \vec{H} = \frac{1}{r}\left(\frac{\partial (r \cdot H_\alpha)}{\partial r}\right) \tag{3.69}$$

Inside of the conductor (see (3.49))

$$H_\alpha(r) = \frac{I}{2 \cdot \pi} \cdot \frac{r}{r_0^2}$$

With (3.69)

$$\text{curl}_z \, \vec{H} = \frac{1}{r}\left[\frac{\partial}{\partial r}\left(r \cdot \frac{I}{2 \cdot \pi} \cdot \frac{r}{r_0^2}\right)\right] = \frac{I}{2 \cdot \pi \cdot r_0^2} \cdot \frac{1}{r}\left(\frac{\partial (r^2)}{\partial r}\right)$$

$$\text{curl}_z \, \vec{H} = \frac{I}{2 \cdot \pi \cdot r_0^2} \cdot \frac{1}{r} \cdot (2 \cdot r)$$

$$\text{curl}_z \vec{H} = \frac{I}{\pi \cdot r_0^2} = J = J_z$$

This result is, as was to be expected, the constant current density in the entire space within the conductor. Outside the conductor (see (3.50)), we have

$$H_\alpha(r) = \frac{I}{2 \cdot \pi} \cdot \frac{1}{r}$$

$$\text{curl}_z \, \vec{H} = \frac{I}{2 \cdot \pi} \cdot \frac{1}{r}\left(\frac{\partial (1)}{\partial r}\right) = 0$$

The magnetic field vectors form concentric circles around the axis of the conductor. In a macroscopic view, the vortices of the magnetic field are therefore easy to recognize (see Fig. 3.2). But the curls of the magnetic field are also present at an infinitesimal surface element. Figure 3.25 depicts a surface element at the position $r = r_1$. It can be approximated by a trapezoid. It is assumed that the surface element is inside the current-carrying conductor. In this case, the path integral of the scalar product of the magnetic field vectors and the path elements of the lateral surfaces of the surface element, divided by the surface element, must be equal to the current density.

The magnetic field forms a vortex with

$$H_\alpha(r_1 + dr) \cdot (r_1 + dr) \cdot d\alpha > H_\alpha(r_1) \cdot r_1 \cdot d\alpha$$

With (3.49), this is confirmed by Fig. 3.25.

The surface A_{Tr} of the surface element is approximated by a trapezoid.

$$A_{\text{Tr}} = \frac{1}{2} \cdot [(r_1 + dr) \cdot d\alpha + r_1 \cdot d\alpha] \cdot dr = \frac{1}{2} \cdot (r_1 \cdot d\alpha + dr \cdot d\alpha + r_1 \cdot d\alpha) \cdot dr$$

The product $(dr \cdot d\alpha)$ in the sum can be neglected. Therefore[16]

$$A_{\text{Tr}} = r_1 \cdot d\alpha \cdot dr \tag{3.70}$$

After (3.49), the magnetic field inside the conductor is

$$H_\alpha(r) = \frac{I}{2 \cdot \pi} \cdot \frac{r}{r_0^2} = K \cdot r$$

K = constant

If $r = r_1$ and $r = (r_1 + dr)$, the path integral of the scalar product of the field vectors and the path elements of the sides of the surface element will be calculated as (see Fig. 3.25)

$$\left(\oint \vec{H} \cdot d\vec{s}\right)_{z=\text{const}} = K \cdot [(r_1 + dr) \cdot d\alpha \cdot (r_1 + dr) - r_1 \cdot d\alpha \cdot r_1]$$
$$= K \cdot \left[r_1^2 \cdot d\alpha + r_1 \cdot dr \cdot d\alpha + r_1 \cdot dr \cdot d\alpha + dr^2 \cdot d\alpha - r_1^2 \cdot d\alpha\right]$$
$$= K \cdot \left[2 \cdot r_1 \cdot dr \cdot d\alpha + dr^2 \cdot d\alpha\right]$$

$\left(dr^2 \cdot d\alpha\right)$ can be neglected

$$\left(\oint \vec{H} \cdot d\vec{s}\right)_{z=\text{const}} = K \cdot [2 \cdot r_1 \cdot dr \cdot d\alpha]$$

$$\text{curl}_z \, \vec{H} = \frac{1}{A_{\text{Tr}}} \cdot \left(\oint \vec{H} \cdot d\vec{s}\right)_{z=\text{const}} = \frac{K \cdot [2 \cdot r_1 \cdot dr \cdot d\alpha]}{r_1 \cdot d\alpha \cdot dr} = \frac{I}{2 \cdot \pi} \cdot \frac{1}{r_0^2} \cdot \frac{2 \cdot r_1 \cdot dr \cdot d\alpha}{r_1 \cdot d\alpha \cdot dr}$$

Thus

$$\text{curl}_z \, \vec{H} = \frac{I}{\pi \cdot r_0^2} = J = J_z$$

3.5.3   Vector Operation $\nabla \times \vec{H}$ in Cartesian Coordinates

In the Cartesian coordinate system, the x-component of the vector $\nabla \times \vec{H}$ is perpendicular to the x-component of the surface element, located in the y–z plane ($x = \text{const}$, see

---

[16]The surface element can therefore be approximated by a rectangle.

Fig. 3.26 x-Component of the vector $\nabla \times \vec{H}$

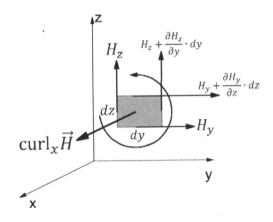

Fig. 3.26). To calculate the component $\text{curl}_x \vec{H}$, the x-component of this surface element must be circumnavigated in the sense of a right-hand screw.

Integrating along the rectangle in the x–y plane yields

$$\left(\oint \vec{H} \cdot d\vec{s}\right)_{x=\text{const}}$$
$$= H_y \cdot dy + \left(H_z + \frac{\partial H_z}{\partial y} \cdot dy\right) \cdot dz - \left(H_y + \frac{\partial H_y}{\partial z} \cdot dz\right) \cdot dy - H_z \cdot dz \tag{3.71}$$

$$\left(\oint \vec{H} \cdot d\vec{s}\right)_{x=\text{const}}$$
$$= H_y \cdot dy + H_z \cdot dz + \frac{\partial H_z}{\partial y} \cdot dy \cdot dz - H_y \cdot dy - \frac{\partial H_y}{\partial z} \cdot dz \cdot dy - H_z \cdot dz$$

Thus

$$\left(\oint \vec{H} \cdot d\vec{s}\right)_{x=\text{const}} = \left(\frac{\partial H_z}{\partial y} - \frac{\partial H_y}{\partial z}\right) \cdot dy \cdot dz$$

According to

(3.58) and (3.59), the x-component of the vector $\text{curl} \, \vec{H} = \nabla \times \vec{H}$ is

$$\text{curl}_x \vec{H} = \frac{1}{dy \cdot dz} \cdot \oint \vec{H} \cdot d\vec{s} = \frac{\partial H_z}{\partial y} - \frac{\partial H_y}{\partial z} \tag{3.72}$$

The other two components of the vector $\nabla \times \vec{H}$ are obtained by cyclically swapping the indices

$$\text{curl}_y \vec{H} = \frac{\partial H_x}{\partial z} - \frac{\partial H_z}{\partial x} \tag{3.73}$$

$$\text{curl}_z \vec{H} = \frac{\partial H_y}{\partial x} - \frac{\partial H_x}{\partial y} \tag{3.74}$$

The three components of curl $\vec{H}$, respectively, $\nabla \times \vec{H}$, can be combined in the form of a matrix

$$\operatorname{curl} \vec{H} = \nabla \times \vec{H} = \begin{vmatrix} \vec{e}_x & \vec{e}_y & \vec{e}_z \\ \frac{\partial}{\partial x} & \frac{\partial}{\partial y} & \frac{\partial}{\partial z} \\ H_x & H_y & H_z \end{vmatrix} \tag{3.75}$$

According to the rules for three-row determinants [7] ($\vec{e}_x, \vec{e}_y, \vec{e}_z$ are the unit vectors)

$$\operatorname{curl} \vec{H} = \nabla \times \vec{H} = \vec{e}_x \left(\frac{\partial H_z}{\partial y} - \frac{\partial H_y}{\partial z} \right) - \vec{e}_y \left(\frac{\partial H_z}{\partial x} - \frac{\partial H_x}{\partial z} \right) + \vec{e}_z \left(\frac{\partial H_y}{\partial x} - \frac{\partial H_x}{\partial y} \right)$$

i.e.,

$$\operatorname{curl} \vec{H} = \nabla \times \vec{H} = \vec{e}_x \left(\frac{\partial H_z}{\partial y} - \frac{\partial H_y}{\partial z} \right) + \vec{e}_y \left(\frac{\partial H_x}{\partial z} - \frac{\partial H_z}{\partial x} \right) + \vec{e}_z \left(\frac{\partial H_y}{\partial x} - \frac{\partial H_x}{\partial y} \right) \tag{3.76}$$

3.5.4 Vector Operation $\nabla \times \vec{H}$ in Spherical Coordinates

3.5.4.1 r-Component of the Vector $\nabla \times \vec{H}$

Integration along the rectangle of the surface $r = $ constant (see Fig. 3.27)

$$\left(\oint \vec{H} \cdot d\vec{s} \right)_{r=\text{const}} = H_\vartheta \cdot r \cdot d\vartheta + \left(H_\alpha + \frac{\partial H_\alpha}{\partial \vartheta} \cdot d\vartheta \right) \cdot r \cdot \sin(\vartheta + d\vartheta) \cdot d\alpha$$
$$- \left(H_\vartheta + \frac{\partial H_\vartheta}{\partial \alpha} \cdot d\alpha \right) \cdot r \cdot d\vartheta - H_\alpha \cdot r \cdot \sin \vartheta \cdot d\alpha \tag{3.77}$$

$$\left(\oint \vec{H} \cdot d\vec{s} \right)_{r=\text{const}} = H_\vartheta \cdot r \cdot d\vartheta + H_\alpha \cdot r \cdot \sin(\vartheta + d\vartheta) \cdot d\alpha$$
$$+ \frac{\partial H_\alpha}{\partial \vartheta} \cdot d\vartheta \cdot r \cdot \sin(\vartheta + d\vartheta) \cdot d\alpha - H_\vartheta \cdot r \cdot d\vartheta$$
$$- \frac{\partial H_\vartheta}{\partial \alpha} \cdot d\alpha \cdot r \cdot d\vartheta - H_\alpha \cdot r \cdot \sin \vartheta \cdot d\alpha$$

With the relationship

$$\sin(x + y) = \sin x \cdot \cos y + \cos x \cdot \sin y$$

we have

$$\left(\oint \vec{H} \cdot d\vec{s} \right)_{r=\text{const}} = H_\alpha \cdot r \cdot (\sin \vartheta \cdot \cos \partial \vartheta + \cos \vartheta \cdot \sin \partial \vartheta) \cdot d\alpha$$
$$+ \frac{\partial H_\alpha}{\partial \vartheta} \cdot d\vartheta \cdot r \cdot (\sin \vartheta \cdot \cos \partial \vartheta + \cos \vartheta \cdot \sin \partial \vartheta) \cdot d\alpha$$
$$- \frac{\partial H_\vartheta}{\partial \alpha} \cdot d\alpha \cdot r \cdot d\vartheta - H_\alpha \cdot r \cdot \sin \vartheta \cdot d\alpha$$

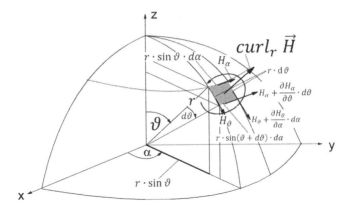

Fig. 3.27 r-Component curl$_r$ $\vec{H}$ of the vector $\nabla \times \vec{H}$ in a spherical coordinate system

$$\left(\oint \vec{H} \cdot d\vec{s}\right)_{r=const} = H_\alpha \cdot r \cdot (\sin\vartheta \cdot \cos\partial\vartheta) \cdot d\alpha + H_\alpha \cdot r \cdot (\cos\vartheta \cdot \sin\partial\vartheta) \cdot d\alpha$$
$$+ \frac{\partial}{\partial\vartheta} H_\alpha \cdot d\vartheta \cdot r \cdot (\sin\vartheta \cdot \cos\partial\vartheta) \cdot d\alpha + \frac{\partial H_\alpha}{\partial\vartheta} \cdot d\vartheta \cdot r \cdot (\cos\vartheta \cdot \sin\partial\vartheta) \cdot d\alpha$$
$$- \frac{\partial H_\vartheta}{\partial\alpha} \cdot d\alpha \cdot r \cdot d\vartheta - H_\alpha \cdot r \cdot \sin\vartheta \cdot d\alpha$$

Since $(\cos\partial\vartheta)$ approaches 1 for small values of $\partial\vartheta$, both marked expressions are equal. In addition, the sine function for small angles can be replaced by its argument, i.e., $(\sin\partial\vartheta \to \partial\vartheta)$

$$\left(\oint \vec{H} \cdot d\vec{s}\right)_{r=const} =$$
$$+ H_\alpha \cdot r \cdot \cos\vartheta \cdot d\vartheta \cdot d\alpha + \frac{\partial H_\alpha}{\partial\vartheta} \cdot d\vartheta \cdot r \cdot (\sin\vartheta \cdot 1) \cdot d\alpha \qquad (3.78)$$
$$+ \frac{\partial H_\alpha}{\partial\vartheta} \cdot d\vartheta \cdot r \cdot (\cos\vartheta \cdot \partial\vartheta) \cdot d\alpha - \frac{\partial H_\vartheta}{\partial\alpha} \cdot d\alpha \cdot r \cdot d\vartheta$$

The expression

$$\frac{\partial H_\alpha}{\partial\vartheta} \cdot d\vartheta \cdot r \cdot \cos\vartheta \cdot \partial\vartheta \cdot d\alpha$$

can be neglected in comparison to

$$H_\alpha \cdot r \cdot \cos\vartheta \cdot d\vartheta \cdot d\alpha$$

Thus (3.78) yields

$$\left(\oint \vec{H} \cdot d\vec{s}\right)_{r=const} = H_\alpha \cdot r \cdot \cos\vartheta \cdot d\vartheta \cdot d\alpha + \frac{\partial H_\alpha}{\partial\vartheta} \cdot d\vartheta \cdot r \cdot \sin\vartheta \cdot d\alpha - \frac{\partial H_\vartheta}{\partial\alpha} \cdot d\alpha \cdot r \cdot d\vartheta$$
$$\left(\oint \vec{H} \cdot d\vec{s}\right)_{r=const} = \left(H_\alpha \cdot \cos\vartheta + \frac{\partial H_\alpha}{\partial\vartheta} \cdot \sin\vartheta - \frac{\partial H_\vartheta}{\partial\alpha}\right) \cdot r \cdot d\vartheta \cdot d\alpha$$
$$\left(\oint \vec{H} \cdot d\vec{s}\right)_{r=const} = \left(\frac{\partial}{\partial\vartheta}(\sin\vartheta \cdot H_\alpha) - \frac{\partial H_\vartheta}{\partial\alpha}\right) \cdot r \cdot d\vartheta \cdot d\alpha$$

Since $d\vartheta$ is infinitesimal the trapezoidal, shaded in Fig. 3.27, can be approximated by a rectangle. With
(3.58) and (3.59) for the r-component, we have

$$
\begin{aligned}
\operatorname{curl}_r \vec{H} &= \frac{1}{(r \cdot \sin \vartheta \cdot d\alpha) \cdot (r \cdot d\vartheta)} \cdot \left(\oint \vec{H} \cdot d\vec{s} \right)_{r=\text{const}} \\
&= \frac{1}{(r \cdot \sin \vartheta \cdot d\alpha) \cdot (r \cdot d\vartheta) \cdot \left(\frac{\partial}{\partial \vartheta}(\sin \vartheta \cdot H_\alpha) - \frac{\partial H_\vartheta}{\partial \alpha} \right) \cdot r \cdot d\vartheta \cdot d\alpha}
\end{aligned}
$$

$$
\operatorname{curl}_r \vec{H} = \frac{1}{r \cdot \sin \vartheta} \cdot \left(\frac{\partial}{\partial \vartheta}(\sin \vartheta \cdot H_\alpha) - \frac{\partial H_\vartheta}{\partial \alpha} \right) \tag{3.79}
$$

3.5.4.2 ϑ-Component of the Vector $\nabla \times \vec{H}$

Integration along the rectangle in the surface $\vartheta = $ constant (Fig. 3.28)

$$
\begin{aligned}
\left(\oint \vec{H} \cdot d\vec{s} \right)_{\vartheta=\text{const}} &= H_\alpha \cdot r \cdot \sin \vartheta \cdot d\alpha + \left(H_r + \frac{\partial H_r}{\partial \alpha} \cdot d\alpha \right) \cdot dr \\
&\quad - \left(H_\alpha + \frac{\partial H_\alpha}{\partial r} \cdot dr \right) \cdot (r + dr) \cdot \sin \vartheta \cdot d\alpha \tag{3.80} \\
&\quad - H_r \cdot dr
\end{aligned}
$$

$$
\begin{aligned}
\left(\oint \vec{H} \cdot d\vec{s} \right)_{\vartheta=\text{const}} &= H_\alpha \cdot r \cdot \sin \vartheta \cdot d\alpha + H_r \cdot dr + \frac{\partial H_r}{\partial \alpha} \cdot d\alpha \cdot dr \\
&\quad - H_\alpha \cdot (r + dr) \cdot \sin \vartheta \cdot d\alpha - \frac{\partial H_\alpha}{\partial r} \cdot dr \cdot (r + dr) \cdot \sin \vartheta \cdot d\alpha \\
&\quad - H_r \cdot dr
\end{aligned}
$$

$\left(\frac{\partial H_\alpha}{\partial r} \cdot dr \cdot dr \cdot \sin \vartheta \cdot d\alpha \right)$ is negligible in relation to $\left(\frac{\partial H_\alpha}{\partial r} \cdot r \cdot dr \cdot \sin \vartheta \cdot d\alpha \right)$.

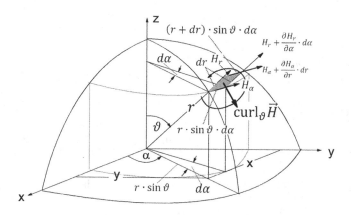

Fig. 3.28 ϑ-Component of the vector $\nabla \times \vec{H}$ in the spherical coordinate system

Thus

$$\left(\oint \vec{H} \cdot d\vec{s}\right)_{\vartheta=\text{const}} = \frac{\partial H_r}{\partial \alpha} \cdot d\alpha \cdot dr - H_\alpha \cdot dr \cdot \sin \vartheta \cdot d\alpha - \frac{\partial H_\alpha}{\partial r} \cdot dr \cdot r \cdot \sin \vartheta \cdot d\alpha$$

According to (3.58) and (3.59), the ϑ-component is

$$\text{curl}_\vartheta \vec{H} = \frac{1}{r \cdot \sin \vartheta \cdot d\alpha \cdot dr} \cdot \left(\oint \vec{H} \cdot d\vec{s}\right)_{\vartheta=\text{const}}$$

$$= \frac{1}{r \cdot \sin \vartheta} \cdot \frac{\partial H_r}{\partial \alpha} - \frac{1}{r} H_\alpha - \frac{\partial H_\alpha}{\partial r}$$

$$= \frac{1}{r \cdot \sin \vartheta} \cdot \left(\frac{\partial H_r}{\partial \alpha} - H_\alpha \cdot \sin \vartheta - \frac{\partial H_\alpha}{\partial r} \cdot r \cdot \sin \vartheta\right)$$

Thus

$$\text{curl}_\vartheta \vec{H} = \frac{1}{r \cdot \sin \vartheta} \cdot \left(\frac{\partial H_r}{\partial \alpha} - \frac{\partial}{\partial r}(r \cdot \sin \vartheta \cdot H_\alpha)\right) \tag{3.81}$$

3.5.4.3 α-Component of the Vector $\nabla \times \vec{H}$

Integration along the path around the rectangle in the surface $\alpha=$ constant (Fig. 3.29)

$$\left(\oint \vec{H} \cdot d\vec{s}\right)_{\alpha=\text{const}} = H_r \cdot dr + \left(H_\vartheta + \frac{\partial H_\vartheta}{\partial r} \cdot dr\right) \cdot (r + dr) \cdot d\vartheta$$

$$- \left(H_r + \frac{\partial H_r}{\partial \vartheta} \cdot d\vartheta\right) \cdot dr - H_\vartheta \cdot r \cdot d\vartheta \tag{3.82}$$

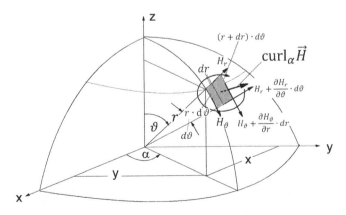

Fig. 3.29 α-Component of the vector $\nabla \times \vec{H}$ in the spherical coordinate system

$$\left(\oint \vec{H} \cdot d\vec{s}\right)_{\alpha=const} = H_r \cdot dr$$

$$+ H_\vartheta \cdot r \cdot d\vartheta + H_\vartheta \cdot dr \cdot d\vartheta + \frac{\partial H_\vartheta}{\partial r} \cdot dr \cdot r \cdot d\vartheta + \frac{\partial H_\vartheta}{\partial r} \cdot dr \cdot dr \cdot d\vartheta$$

$$- H_r \cdot dr - \frac{\partial H_r}{\partial \vartheta} \cdot d\vartheta \cdot dr - H_\delta \cdot r \cdot d\vartheta$$

The expression

$$\frac{\partial H_\vartheta}{\partial r} \cdot dr \cdot dr \cdot d\vartheta$$

can be neglected in comparison to

$$\frac{\partial H_\vartheta}{\partial r} \cdot dr \cdot r \cdot d\vartheta$$

Thus

$$\left(\oint \vec{H} \cdot d\vec{s}\right)_{\alpha=const} = H_\vartheta \cdot dr \cdot d\vartheta + \frac{\partial H_\vartheta}{\partial r} \cdot dr \cdot r \cdot d\vartheta - \frac{\partial H_r}{\partial \vartheta} \cdot d\vartheta \cdot dr$$

$$\text{curl}_\alpha \vec{H} = \frac{1}{r \cdot d\vartheta \cdot dr} \cdot \oint \vec{H} \cdot d\vec{s} = \frac{1}{r \cdot d\vartheta \cdot dr} \cdot \left(H_\vartheta + r \cdot \frac{\partial H_\vartheta}{\partial r} - \frac{\partial H_r}{\partial \vartheta}\right) \cdot dr \cdot d\vartheta$$

and

$$\text{curl}_\alpha \vec{H} = \frac{1}{r}\left(H_\vartheta + r \cdot \frac{\partial H_\vartheta}{\partial r} - \frac{\partial H_r}{\partial \vartheta}\right)$$

or

$$\text{curl}_\alpha \vec{H} = \frac{1}{r} \cdot \left(\frac{\partial}{\partial r}(r \cdot H_\vartheta) - \frac{\partial H_r}{\partial \vartheta}\right) \tag{3.83}$$

The three components of $\nabla \times \vec{H}$ can be combined in the form of a matrix

$$\nabla \times \vec{H} = \frac{1}{r^2 \cdot \sin\vartheta}\begin{vmatrix} \vec{e}_r & r \cdot \vec{e}_\vartheta & r \cdot \sin\vartheta \cdot \vec{e}_\alpha \\ \frac{\partial}{\partial r} & \frac{\partial}{\partial \vartheta} & \frac{\partial}{\partial \alpha} \\ H_r & r \cdot H_\vartheta & r \cdot \sin\vartheta \cdot H_\alpha \end{vmatrix} \tag{3.84}$$

According to the rules for three-row determinants, ([7], $\vec{e}_x, \vec{e}_y, \vec{e}_z$ = unit vectors)

$$\nabla \times \vec{H} = \frac{1}{r^2 \cdot \sin\vartheta} \cdot \left[\vec{e}_r\left(\frac{\partial(r \cdot \sin\vartheta \cdot H_\alpha)}{\partial \vartheta} - \frac{\partial(r \cdot H_\vartheta)}{\partial \alpha}\right)\right.$$

$$\left. - r \cdot \vec{e}_\vartheta\left(\frac{\partial(r \cdot \sin\vartheta \cdot H_\alpha)}{\partial r} - \frac{\partial H_r}{\partial \alpha}\right) + \vec{e}_\alpha \cdot (r \cdot \sin\vartheta \cdot)\left(\frac{\partial(r \cdot H_\vartheta)}{\partial r} - \frac{\partial H_r}{\partial \vartheta}\right)\right]$$

i.e.,

$$\nabla \times \vec{H} = \frac{1}{r \cdot \sin \vartheta} \cdot \left(\frac{\partial (\sin \vartheta \cdot H_\alpha)}{\partial \vartheta} - \frac{\partial H_\vartheta}{\partial \alpha} \right) \vec{e}_r$$
$$+ \frac{1}{r} \left(\frac{1}{\sin \vartheta} \cdot \frac{\partial H_r}{\partial \alpha} - \frac{\partial (r \cdot H_\alpha)}{\partial r} \right) \vec{e}_\vartheta + \frac{1}{r} \cdot \left(\frac{\partial (r \cdot H_\vartheta)}{\partial r} - \frac{\partial H_r}{\partial \vartheta} \right) \vec{e}_\alpha \qquad (3.85)$$

3.5.5   Vector Analysis Formulas

In this section, some formulas of vector analysis are summarized. These will be required in the following sections. They can be derived from the rules of differential calculus.

$$\text{curl grad } \varphi = \nabla \times \nabla \varphi = 0 \qquad (3.86)$$

$$\text{div curl } \vec{V} = \nabla \cdot \left(\nabla \times \vec{V} \right) = 0 \qquad (3.87)$$

$$\text{curl} \left(\text{curl } \vec{V} \right) = \text{grad div } \vec{V} - \nabla^2 \vec{V}$$
or
$$\nabla \times \left(\nabla \times \vec{V} \right) = \nabla \left(\nabla \cdot \vec{V} \right) - \nabla^2 \vec{V} \qquad (3.88)$$

$$\text{div} \left(\vec{V} \times \vec{B} \right) = \vec{B} \cdot \text{curl } \vec{V} - \vec{V} \cdot \text{curl } \vec{B}$$
or
$$\nabla \cdot \left(\vec{V} \times \vec{B} \right) = \vec{B} \cdot \left(\nabla \times \vec{V} \right) - \vec{V} \cdot \left(\nabla \times \vec{B} \right) \qquad (3.89)$$

The double application of the Nabla operator ∇^2 to a vector field (see (2.48)) is explained in Cartesian coordinates as follows:

$$\nabla^2 \vec{V} = \left(\nabla^2 V_x \right) \cdot \vec{e}_x + \left(\nabla^2 V_y \right) \cdot \vec{e}_y + \left(\nabla^2 V_z \right) \cdot \vec{e}_z \qquad (3.90)$$

3.6   Magnetic Vector Potential

The introduction of potentials proves to be useful, among other things, for solving Maxwell's equations. Potentials are used to decouple these equations and make them easier to solve. They are mathematical auxiliary quantities which, in contrast to magnetic and electric fields, do not correspond to reality in the physical sense. While the scalar potential, as explained in Sect. 3.3, can only be applied to surfaces where the current density is zero, the magnetic vector potential is important for the calculation of fields inside current-carrying conductors but also for wave propagation problems where the magnetic effect of displacement currents[17] has an influence. The starting point for the introduction of the magnetic vector potential is (3.59)

$$\nabla \times \vec{H} = \vec{J}$$

or

$$\nabla \times \vec{B} = \mu \cdot \vec{J}$$

Between the current I_{con} concatenated with a surface $\vec{A}$ and the current density $\vec{J}$, the following relationship applies:

$$I_{con} = \iint\limits_A \vec{J} \cdot d\vec{A} \tag{3.91}$$

With (3.60) and (3.91), we have

$$I_{con} = \iint\limits_A \nabla \times \vec{H} \cdot d\vec{A} \tag{3.92}$$

After (3.45), the integral along the closed contour C is equal to the current I_{con} flowing into an area defined by this contour

$$\oint\limits_C \vec{H} \cdot d\vec{s} = I_{con}$$

From (3.92), we obtain

$$\oint\limits_C \vec{H} \cdot d\vec{s} = \iint\limits_A \nabla \times \vec{H} \cdot d\vec{A} \tag{3.93}$$

Equation (3.93) is called the "Stokes' theorem"[18]. It applies generally to all vector fields.
   Statement of the Stokes' theorem:

The integral of a vector field along a contour C is equal to the integral of the curl of this vector field executed over the surface A formed by the contour C.

---

[17]For displacement current, see Sect. 4.2.

[18]Sir Stokes, Gabriel, British physicist and mathematician, *1819, †1903.

In contrast to the static electric field, the magnetic field is source free. The magnetic field lines are closed. Consequently, there are no magnetic charges as sources of the magnetic field lines, therefore

$$\nabla \cdot \vec{H} = 0 \tag{3.94}$$

With (3.87)

$$\nabla \cdot \left(\nabla \times \vec{A} \right) = 0$$

a new vector field $\vec{A}$ is introduced[19]. With this relationship and (3.94), we have

$$\nabla \cdot \vec{H} = \nabla \cdot \left(\nabla \times \vec{A} \right) = 0$$

or

$$\nabla \cdot \vec{B} = \mu \left[\nabla \cdot \left(\nabla \times \vec{A} \right) \right] = 0 \tag{3.95}$$

The vector field $\vec{A}$ is called magnetic vector potential. According to (3.95), the magnetic flux density $\vec{B}$ and the vector potential $\vec{A}$ are related by

$$\nabla \cdot \vec{B} = \nabla \cdot \left(\nabla \times \vec{A} \right) = 0$$

and consequently

$$\nabla \times \vec{A} = \vec{B} \tag{3.96}$$

With Ampère's law (3.59)

$$\nabla \times \vec{H} = \vec{J}$$

the relationship between vector potential $\vec{A}$ and current density $\vec{J}$ is obtained

$$\nabla \times \left(\nabla \times \vec{A} \right) = \mu \cdot \vec{J} \tag{3.97}$$

With (3.88)

$$\nabla \times \left(\nabla \times \vec{V} \right) = \nabla \left(\nabla \cdot \vec{V} \right) - \nabla^2 \vec{V}$$

we get

$$\mu \cdot \vec{J} = \nabla \left(\nabla \cdot \vec{A} \right) - \nabla^2 \vec{A} \tag{3.98}$$

---

[19]For the vector potential, the same designation $\vec{A}$ is used as for the surface vector $\vec{A}$. From the context, it is always clear what is the meaning of $\vec{A}$. There will be no confusion.

Fig. 3.30 Vector potential of
a space containing a location
variable current density

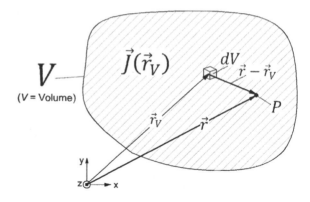

Equation (3.98) contains the expression $\left(\nabla \cdot \vec{A} \right)$. The divergence of the vector potential
has not yet been defined and can be freely selected.

Setting

$$\nabla \cdot \vec{A} = 0 \tag{3.99}$$

(3.98) is simplified to

$$\nabla^2 \vec{A} = -\mu \cdot \vec{J} \tag{3.100}$$

Thus components of the magnetic vector potential $\vec{A}$, expressed with current density
components, are

$$
\begin{aligned}
\nabla^2 A_x &= -\mu \cdot J_x \\
\nabla^2 A_y &= -\mu \cdot J_y \\
\nabla^2 A_z &= -\mu \cdot J_z
\end{aligned}
\tag{3.101}
$$

Equation (3.100) is the potential equation of the magnetic vector potential for a space
with current density. It is the analogous relationship to the Poisson Eq. (2.48)

$$\nabla^2 \varphi = -\frac{\varrho}{\varepsilon}$$

Each of the three component equations in (3.101) is an analogous to Eq. (2.48), whereby
instead of the quotient ϱ/ε the corresponding component of the current density multiplied by the permeability μ in this equation.

If the charge density ϱ and the permittivity ε are constant in space, then (2.36) gives
the electrical scalar potential (see also Fig. 2.17)

(2.36)

$$\varphi(\vec{r}) = \frac{1}{4 \cdot \pi \cdot \varepsilon} \iiint\limits_V \varrho(\vec{r}_V) \cdot \frac{dV}{|\vec{r} - \vec{r}_V|}$$

From the analogy of (3.100) with (2.48), one gets with (2.36) and the analog conclusion for the components of the vector potential in the Cartesian coordinate system (see Fig. 3.30)

$$A_x = \frac{\mu}{4 \cdot \pi} \iiint\limits_V J_x(\vec{r}_V) \cdot \frac{\mathrm{d}V}{|\vec{r} - \vec{r}_V|}$$

$$A_y = \frac{\mu}{4 \cdot \pi} \iiint\limits_V J_y(\vec{r}_V) \cdot \frac{\mathrm{d}V}{|\vec{r} - \vec{r}_V|} \qquad (3.102)$$

$$A_z = \frac{\mu}{4 \cdot \pi} \iiint\limits_V J_z(\vec{r}_V) \cdot \frac{\mathrm{d}V}{|\vec{r} - \vec{r}_V|}$$

Equation (3.102) is the solution of (3.100). The integration of the current density must be carried out over the current density in the entire volume V, considering all current-carrying conductors in this volume. Equations (3.102) can be summarized as

$$\vec{A}(\vec{r}) = \frac{\mu}{4 \cdot \pi} \iiint\limits_V \frac{\vec{J}(\vec{r}_V)}{|\vec{r} - \vec{r}_V|} \mathrm{d}V \qquad (3.103)$$

Equation (3.103) fulfills Eq. (3.99), as can be proven[20], i.e.,

$$\nabla \cdot \vec{A} = 0 \qquad (3.104)$$

3.7 Law of Biot-Savart

The law of Biot and Savart[21] specifies how to calculate the magnetic flux density produced in a point P of a space by a conductor loop carrying a current I. The starting point for the derivation of this law is a current I flowing through a conductor section with a very small cross section A and a short length $\mathrm{d}s$. Such a conductor section is called an elementary conductor. One can imagine such an elementary conductor as an infinitesimally small section of a conductor loop. In Fig. 3.31, it is assumed that the elementary conductor is situated in the origin of the Cartesian coordinate system and oriented in the x–y plane in the positive x-direction.

If A is the cross section of the elementary conductor, $\mathrm{d}\vec{s}$ the vector of the length element, and I the current in the conductor, the vector of the current density $\vec{J}$ will be

$$\vec{J} = \frac{I}{A} \cdot \frac{\mathrm{d}\vec{s}}{\mathrm{d}s} \qquad (3.105)$$

---

[20]See [3] p. 197.

[21]Biot, Jean-Baptiste, French physicist and astronomer, *1774, †1862

Savart, Félix, French physicist, *1791, †1841.

Fig. 3.31 Current-carrying
elementary conductor

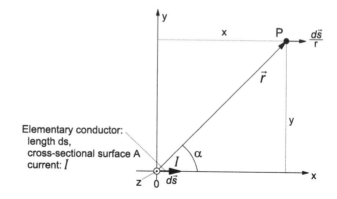

The volume element dV of the elementary conductor (A = cross-sectional surface of the elementary conductor) is

$$dV = A \cdot ds \tag{3.106}$$

The magnetic field $d\vec{H}$ generated in space by the current of the elementary conductor can be calculated with (3.103) in conjunction with (3.96) and (3.46)

$$\vec{H} = \frac{\nabla \times \vec{A}}{\mu} = \frac{\vec{B}}{\mu}$$

Since the elementary conductor is located in the origin of the coordinate system follows $\vec{r}_V = 0$.

$$d\vec{H} = \frac{1}{4 \cdot \pi} \left[\nabla \times \left(\iiint\limits_V \frac{\vec{J}}{r} dv \right) \right] = \frac{1}{4 \cdot \pi} \cdot \left[\nabla \times \left(\int\limits_0^{ds} \frac{I}{A} \cdot \frac{d\vec{s}}{ds} \cdot \frac{1}{r} \cdot A \cdot ds \right) \right]$$

or

$$d\vec{H} = \frac{I}{4 \cdot \pi} \left[\nabla \times \left(\frac{d\vec{s}}{r} \right) \right] \tag{3.107}$$

The vector $d\vec{s}/r$ in (3.107) has only a component in x-direction (see Fig. 3.31) which is independent of the z-coordinate. Thus, the x- and y-components of the operation $\left[\nabla \times \left(\frac{d\vec{s}}{r} \right) \right]$ are zero according to Eqs. (3.72) and (3.73)

$$\mathrm{curl}_x \left(\frac{ds}{r} \right) = 0$$

and

$$\mathrm{curl}_y \left(\frac{ds}{r} \right) = 0$$

Fig. 3.32 Biot-Savart, magnetic field at point P

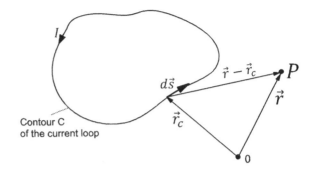

Contour C
of the current loop

According to (3.74), the z-component is

$$\mathrm{curl}_z\left(\frac{\mathrm{d}s}{r}\right) = -\frac{\partial}{\partial y}\left(\frac{\mathrm{d}s}{r}\right) = -\frac{\partial}{\partial y}\left(\frac{\mathrm{d}s}{\sqrt{x^2+y^2}}\right) = -\frac{\partial}{\partial y}\left(\mathrm{d}s\cdot(x^2+y^2)^{-\frac{1}{2}}\right)$$

$$\mathrm{curl}_z\left(\frac{\mathrm{d}s}{r}\right) = -\mathrm{d}s\cdot\left(-\frac{1}{2}\cdot(x^2+y^2)^{-\frac{3}{2}}\cdot 2\cdot y\right) = \frac{y\cdot\mathrm{d}s}{\left(\sqrt{x^2+y^2}\right)^3}$$

$$\mathrm{curl}_z\left(\frac{\mathrm{d}s}{r}\right) = \frac{y\cdot\mathrm{d}s}{r^3} = \frac{r\cdot\sin\alpha\cdot\mathrm{d}s}{r^3}$$

Consequently (see Fig. 3.6 and (3.12))

$$\nabla\times\left(\frac{\mathrm{d}\vec{s}}{r}\right) = \frac{\mathrm{d}\vec{s}\times\vec{r}}{r^3}\tag{3.108}$$

Thus (3.107) simplifies to

$$\mathrm{d}\vec{H} = \frac{I}{4\cdot\pi}\cdot\frac{\mathrm{d}\vec{s}\times\vec{r}}{r^3}\tag{3.109}$$

The magnetic field in point P is determined by the current and by the vector product of the path element $\mathrm{d}\vec{s}$ with the vector $\vec{r}/r^3$ oriented to this point P. The magnitude of the magnetic field is thus proportional to $\vec{r}/r^3$, i.e., inversely proportional to r^2 or inversely proportional to the square of the distance between point P and the current element.

In order to calculate the magnetic field at point P generated by the total current flowing through the closed loop where the origin of the coordinate system is arbitrarily selected, (3.109) must be modified accordingly (see Fig. 3.32). First, the integration has to be performed along the entire current loop forming the contour C. In addition, the vector $(\vec{r}-\vec{r}_C)$ replaces vector $\vec{r}$ in the numerator of (3.110) and the absolute value $|\vec{r}-\vec{r}_C|$ in the denominator replaces the distance r:

$$\vec{H}(\vec{r}) = \frac{I}{4\cdot\pi}\cdot\oint_C\frac{\mathrm{d}\vec{s}\times(\vec{r}-\vec{r}_C)}{|\vec{r}-\vec{r}_C|^3}\tag{3.110}$$

Equation (3.110) is called the law of Biot-Savart.

Time-Varying Electric and Magnetic Fields

4

The previous three chapters dealt with time-independent electric and magnetic fields. As we will see in this chapter, time-varying electric and magnetic fields will generate electromagnetic waves that propagate in space.

Based on Chap. 3, the following section first discusses the switch-on process of an inductor and then derives the equation for the energy density of the magnetic field. Next, the focus is on the law of induction, and the second Maxwell's equation, respectively. Based on Chap. 2, the continuity equation is formulated and the displacement current will be discussed. The introduction of the displacement current is Maxwell's achievement. Although it was not possible to prove the existence of the magnetic field, due to the displacement current with the measuring equipment available to him, he postulated the existence of the displacement current. The introduction of the displacement current was necessary, because without the displacement current Ampére's law for an open alternating current circuit has no validity.

4.1 Inductance

Inductors[1] in various designs are used in electrical circuits for a variety of tasks, e.g., storing energy, filters, etc. Figure 4.1 shows a simple circuit, where an inductor L and a resistor R are connected in series with a DC voltage source V_0 and a switch S. The time function of the current $i(t)$ can be displayed with an oscilloscope due to Ohm's law

$$i(t) = \frac{v_r(t)}{R}$$

---

[1]Also called coil, choke, or reactor.

© Springer Fachmedien Wiesbaden GmbH, part of Springer Nature 2020
J. Donnevert, *Maxwell´s Equations*, https://doi.org/10.1007/978-3-658-29376-5_4

Fig. 4.1 Time function of the
current of an inductor

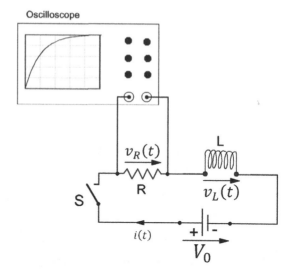

If the switch S is closed at the time $t = 0$, a current $i(t)$ will start to flow. As the measurement shows, the current $i(t)$ does not rise abruptly to its final value. It reaches its final value after some time determined by the ohmic resistance of the circuit, the properties of the inductor, and the voltage V_0 of the voltage source. According to (3.43), the magnetic flux concatenated to the windings is proportional to the current $i(t)$ through the inductor.

$$\Phi_{con}(t) = L \cdot i(t) \qquad (4.1)$$

L is the proportionality factor between the magnetic flux of the inductor and the time-varying current producing it. L is called inductance of the inductor. The value of the inductance depends on the shape of the inductor, the number of windings, and the relative permeability of the material of the core of the inductor. The unit of inductance is Henry[2] (symbol H). (4.1) and (3.6) define the unit of inductance.

$$\text{Unit } (L): \frac{V \cdot s}{A} = H$$

According to (3.22), the voltage v_L at the terminals of the inductor is equal to the temporal change of the magnetic flux concatenated with the windings of the inductor. It is irrelevant what causes the temporal change of the concatenated magnetic flux. In the present case, the temporal change of the concatenated magnetic flux is obviously not caused by a movement of the conductor loops in a magnetic field, but by the temporal change of the electric current $i(t)$ within the windings of the inductor. Since the current $i(t) = 0$ at the

---

[2]Henry, Joseph, American physicist, *1797, †1878.

time of closing the switch S, i.e., at the time $t = 0$, the voltage $v_L(t) = V_0$ must be present at the terminals of the inductance L at $t = 0$. Then, as the measurement shows, the current rises with an exponential function and consequently the voltage at the inductance drops. According to (4.1), the voltage $v_L(t)$ at the terminals of the inductor is

$$v_L(t) = \frac{d\Phi_{con}}{dt} = L \cdot \frac{di(t)}{dt} \tag{4.2}$$

The differential Eq. (4.2) describes the time-varying voltage. Kirchhoff´s mesh rule applied to the mesh in Fig. 4.1 gives

$$V_0 = R \cdot i(t) + L \cdot \frac{di(t)}{dt}$$

Thus

$$dt = \frac{di(t) \cdot L}{V_0 - R \cdot i(t)} = \frac{L}{V_0} \cdot \frac{di(t)}{1 + \left(-\frac{R}{V_0}\right) \cdot i(t)}$$

Integrating both sides of the previous equation

$$\int dt = \frac{L}{V_0} \cdot \int \frac{di(t)}{1 + \left(-\frac{R}{V_0}\right) \cdot i(t)} \tag{4.3}$$

The integral on the right side of this equation is given in [1], page 296, Eq. (2). We have

$$t = -\frac{L}{R} \cdot ln\left(-\frac{R}{V_0} \cdot i(t) + 1\right)$$

or

$$-\frac{R}{L} \cdot t = ln\left(-\frac{R}{V_0} \cdot i(t) + 1\right)$$
$$e^{-\frac{R}{L} \cdot t} = -\frac{R}{V_0} \cdot i(t) + 1$$

Finally, the current $i(t)$ in the circuit and the voltage $v_R(t)$ at the resistor R in Fig. 4.1 are

$$i(t) = \frac{V_0}{R} \cdot \left(1 - e^{-\frac{R}{L} \cdot t}\right) \tag{4.4}$$

$$v_R(t) = V_0 \cdot \left(1 - e^{-\frac{R}{L} \cdot t}\right) \tag{4.5}$$

Figure 4.2 shows the current versus time within the circuit of Fig. 4.1 according to Eq. (4.4).

Fig. 4.2 Time function of the current in the circuit of Fig. 4.1 ($V_0 = 10$ V, $R = 10$ kΩ, $L = 10$ mH)

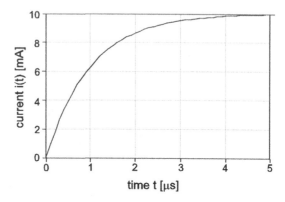

4.2   Energy Density of the Magnetic Field

According to Eq. (2.59), the energy density of the electric field is linked to the field quantities E and D, i.e., to the electric field and the electric flux density. In this section, we describe in which way the energy stored in the magnetic field can be expressed by the magnetic field variables.

In an inductor with windings close together, the magnetic field is concentrated in the core of the inductor (see Fig. 4.3). It is assumed that the permeability μ is independent of

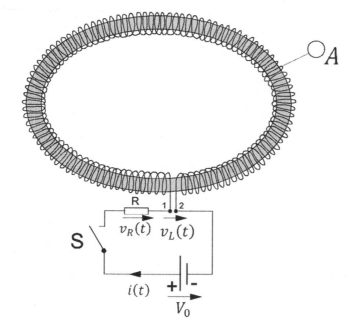

Fig. 4.3 Toroidal inductor

the magnetic field strength. The magnetic flux density B in the cross section A of the core of the inductor can be considered as homogeneous. If the switch S was initially open and is closed at time $t = 0$, according to Fig. 4.2, a time-dependent current $i(t)$ flows and a time-dependent voltage is generated at the terminals 1 and 2. The electrical energy supplied to the magnetic field of the toroidal coil at terminals 1 and 2 up to time t_0 is

$$W = \int_0^{t_0} i(t) \cdot v_L(t) \, dt \tag{4.6}$$

The magnetic flux in the core of the inductor, concatenated with the N windings (see 3.20 and 3.38), is

$$\Phi_{con}(t) = N \cdot A \cdot B(t) \tag{4.7}$$

and according to (3.22)

$$v_L(t) = \frac{d\Phi_{con}}{dt} = N \cdot A \cdot \frac{dB}{dt} \tag{4.8}$$

$N = $ number of windings.

According to Ampère's law (see 3.43 and 3.45), the current I_{con} concatenated with the ring inductor is equal to the length l of the field lines in the core of the ring inductor multiplied by the magnetic field H present in the core at time t

$$I_{con}(t) = i(t) \cdot N = l \cdot H(t) \tag{4.9}$$

Consequently

$$i(t) = \frac{l}{N} \cdot H(t) = \frac{l}{N} \cdot \frac{B(t)}{\mu} \tag{4.10}$$

With (4.6), (4.8), and (4.10), we get

$$W = \int_0^{t_0} \frac{l}{N} \cdot \frac{B(t)}{\mu} \cdot N \cdot A \cdot \frac{dB}{dt} dt \tag{4.11}$$

In this equation, the product $l \cdot A$ is equal to the volume V of the core of the ring inductor. Therefore

$$W = \frac{V}{\mu} \int_0^{t_0} B(t) \cdot \frac{dB}{dt} dt \tag{4.12}$$

The following applies to the product under the integral:

$$B(t) \cdot \frac{dB}{dt} = \frac{d}{dt} \left[\frac{1}{2} \cdot B(t)^2 \right] \tag{4.13}$$

We can verify this equation using the chain rule of differential calculus.

Therefore, with $B(t = 0) = 0$ and $(B(t_0) = B_{t_0}$

$$W = \frac{V}{\mu}\left[\frac{1}{2}B(t)^2\right]_0^{t_0} = \frac{V}{\mu}\cdot\frac{B_{t_0}^2}{2} = \frac{V}{\mu}\cdot\frac{B^2}{2} \tag{4.14}$$

The upper limit $B_{t_0} = B$ is the magnetic flux density at the at time t_0. The energy density w_{magn} of a homogeneous magnetic field in the core of the inductor according to (4.14) therefore is

$$w_{\text{magn}} = \frac{1}{\mu}\cdot\frac{B^2}{2} = \frac{1}{2}B\cdot H = \frac{1}{2}\cdot\mu\cdot H^2 \tag{4.15}$$

4.3   Faraday's Law of Induction

Sections 3.1.1 and 3.1.2 of the previous chapter were demonstrated that an electric field or a voltage V_{12} is induced in a conductor loop moving in a static magnetic field. The prerequisite for this is that the magnetic flux Φ_{con}, concatenated to the loop, changes with time as a result of a change in shape or of the movement of the loop (see Figs. 3.9 and 3.10). Equation (4.16) describes it (see 3.22)

$$V_{12} = -\frac{d\Phi_{\text{con}}}{dt} \tag{4.16}$$

But even in a static conductor loop, the contour of which does not change with time, an electric field will be induced if the magnetic flux, concatenated to the conductor loop, changes with time.

To derive the law of induction, let us look at Fig. 4.4. In this figure, a concatenated magnetic flux $\Phi_{\text{con}}(t)$, changing with time, is created without moving or changing the

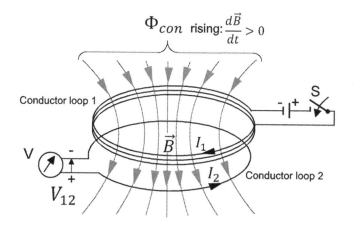

Fig. 4.4 Switching on process in conductor loop 1: Induction of a voltage V_{12} and a current I_2 in the conductor loop 2, concatenated with the conductor loop 1 (The magnetic field lines penetrating both conductor loops are enclosed in itself. Only a section is shown in the figure)

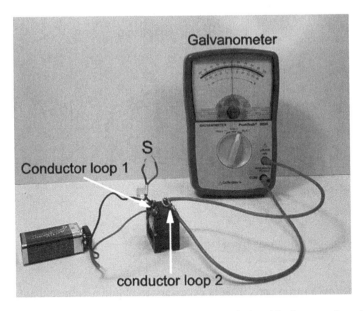

Fig. 4.5 Measuring arrangement for measuring the voltage induced in the second conductor loop

shape of the loop[3]. The figure shows two closely adjacent conductor loops. A voltage source is connected to the conductor loop 1 via a switch S. As soon as switch S is closed the current begins to flow. In a short period of time, the current increases from 0 A until it reaches its maximum (see Fig. 4.2). The maximum is determined by Ohm's resistance of the loop and the voltage of the voltage source. After the maximum has been reached, the current in the conductor loop 1 will be constant[4]. During the switch-on process, a time-varying magnetic field is generated by the rising current in conductor loop 1. The magnetic flux density increases

$$\frac{d\vec{B}}{dt} > 0$$

The magnetic flux Φ_{con} concatenated with the conductor loop 1 is also concatenated with conductor loop 2. As a result, a voltage V_{12} is generated at the terminals of loop 2 during the switch-on process and a current I_2 flows in conductor loop 2 in the indicated direction. After the switch-on process is completed, the voltage V_{12} drops to 0 V and the current flow in the conductor loop 2 ends. The current I_2 is called the induced current, and the voltage V_{12} is referred to as induction voltage.

With a simple electrical circuit, as shown in Fig. 4.5, the induced voltage can be verified. In the circuit, the conductor loops 1 and 2 consist of a large number of windings

---

[3]The arrangement of the upper loop in Fig. 4.4 corresponds to the circuit in Fig. 4.1.

[4]The current loop 1, as well as the current loop 2, has a finite inductance and a finite Ohmic resistance. The exact time function of the current intensity after closing switch S was derived in Sect. 4.1.

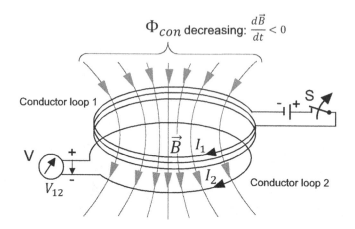

Fig. 4.6 Switching off the current in conductor loop 1

wound together on an iron core. This is necessary to obtain a voltage that can be meas-ured with a simple galvanometer. The iron core ensures that the magnetic flux is concen-trated within the windings. On the left hand, we see a 9 V battery. One pole of the battery is connected to winding 1 by a wire. The other pole of the battery is connected to the other end of winding 1 by the switch S realized by two bare wire ends.

As soon as the switch S is closed, a voltage of about 10 mV can be read on the galva-nometer during the switch-on process. After a short time, when the switch-on procedure is finished, the display drops back to 0 V.

The direction of current I_2 in the conductor loop 2 of Fig. 4.4 is opposite to the direc-tion of the current I_1 in the conductor loop 1. Because of the law of conservation of energy, the magnetic flux generated by current I_2 must counteract the magnetic flux gen-erated by current I_1. This is referred to as Lenz's law.[5]

In Fig. 4.6, the same two loops are shown as in Fig. 4.4 but this time the switch S will be opened. The current in conductor loop 2 does not stop abruptly but as a result of the resistance and the inductance of the conductor, loop decreases with an exponential func-tion. During the switch-off process, the magnetic flux density decreases with the current intensity

$$\frac{d\vec{B}}{dt} < 0$$

As a result, the induced voltage reverses its polarity with respect to the switch-on pro-cess, and the current in loop 2 also changes its direction. In the measuring arrange-ment shown in Fig. 4.5, after opening the switch, the measuring instrument first shows

---

[5]Lenz, Heinrich, Friedrich, Emil, German-Russian physicist, *1804, †1865.

Fig. 4.7 Temporal change
of the magnetic flux in the
infinitesimal surface element
dA

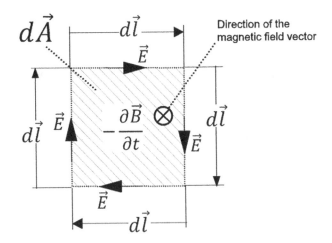

a voltage of -10 mV which gradually approaches 0 V. When switching off, the current direction is the same in both loops. Clearly formulated, according to the Lenz's law: "The current in conductor loop 2 counteracts the change in flux density."

The cause of the temporal change of the concatenated magnetic flux is irrelevant. For generating induced voltage V_{12}, it is only important that the concatenated magnetic flux changes with time.

At this point, there is the question of whether the occurrence of an electric field $\vec{E}$ generating an electric voltage requires the presence of a conductor loop or whether a conductor loop is only required for the verification of the induced electric field. In fact, the conductor loop is not required for the occurrence of the electric field. Every temporal change in the magnetic flux density, also in free space, is "accompanied" by an electric field. Maxwell[6] has not been able to prove this when formulating the law of induction. Heinrich Hertz[7] confirmed this fact by the discovery of the "radio waves" in 1886 at the Technical University of Karlsruhe.

Of course, a change in the magnetic field over time at a location, where there is no current-carrying conductor, must have a cause. This cause can be, for example, a conductor located at a certain distance from the location under consideration and a current changing with time. In Chap. 5, such excitation of an electromagnetic field will be discussed in detail using the example of the Hertzian dipole.

To derive the relationship between a changing magnetic flux and the electric field, we first consider an infinitesimal surface element $d\vec{A}$ (see Fig. 4.7), a square with the side $d\vec{l}$. A decreasing magnetic flux density with time ($-\partial\vec{B}/\partial t < 0$) induces an electric field directed as indicated. No conductor along the border of the surface element $d\vec{A}$ is

---

[6]Maxwell, James Clerk, British physicist, *1831, †1879.

[7]Hertz, Heinrich, German physicist, *1857, †1894.

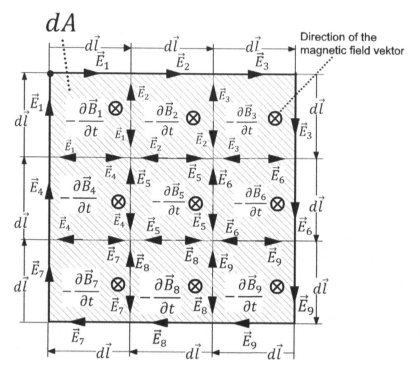

Fig. 4.8 On the law of induction

required. Within the infinitesimal surface element, the rate of change $-\partial \vec{B}/\partial t$ of the flux density $\vec{B}$ can be assumed constant.

After (4.16) with (3.14 and 3.5), we have

$$4 \cdot \left(\vec{E} \cdot d\vec{l} \right) = -\frac{\partial \Phi_{con}}{\partial t} = -\frac{\partial \vec{B}}{\partial t} \cdot d\vec{A} \tag{4.17}$$

To formulate the law of induction, Eq. (4.17) must be applied to a larger surface. Figure 4.8 shows such a surface which, for reasons of clarity and simplicity, consists of nine infinitesimal surfaces $d\vec{A}$. Note that in general the change in flux density $-\partial \vec{B}/\partial t$ is different in all nine surfaces.

Equation (4.17) is applied analogously to each of the nine surface elements. Starting at the upper left corner of Fig. 4.8, we will have

Circulation 1:

$$+\left(\vec{E}_1 \cdot d\vec{l} \right) - \left(\vec{E}_2 \cdot d\vec{l} \right) + \left(\vec{E}_1 \cdot d\vec{l} \right) - \left(\vec{E}_4 \cdot d\vec{l} \right) + \left(\vec{E}_1 \cdot d\vec{l} \right) + \left(\vec{E}_1 \cdot d\vec{l} \right)$$
$$= -\frac{\partial \vec{B}_1}{\partial t} \cdot d\vec{A}$$

Circulation 2:

$$+\left(\vec{E}_2 \cdot d\vec{l} \right) - \left(\vec{E}_3 \cdot d\vec{l} \right) + \left(\vec{E}_2 \cdot d\vec{l} \right) - \left(\vec{E}_5 \cdot d\vec{l} \right) + \left(\vec{E}_2 \cdot d\vec{l} \right) - \left(\vec{E}_1 \cdot d\vec{l} \right) + \left(\vec{E}_2 \cdot d\vec{l} \right) = -\frac{\partial \vec{B}_2}{\partial t} \cdot d\vec{A}$$

Circulation 3:

$$+\left(\vec{E}_3 \cdot \mathrm{d}\vec{l}\right) + \left(\vec{E}_3 \cdot \mathrm{d}\vec{l}\right) - \left(\vec{E}_6 \cdot \mathrm{d}\vec{l}\right) + \left(\vec{E}_3 \cdot \mathrm{d}\vec{l}\right) - \left(\vec{E}_2 \cdot \mathrm{d}\vec{l}\right) + \left(\vec{E}_3 \cdot \mathrm{d}\vec{l}\right) = -\frac{\partial \vec{B}_3}{\partial t} \cdot \mathrm{d}\vec{A}$$

Circulation 4:

$$-\left(\vec{E}_1 \cdot \mathrm{d}\vec{l}\right) + \left(\vec{E}_4 \cdot \mathrm{d}\vec{l}\right) - \left(\vec{E}_5 \cdot \mathrm{d}\vec{l}\right) + \left(\vec{E}_4 \cdot \mathrm{d}\vec{l}\right) - \left(\vec{E}_7 \cdot \mathrm{d}\vec{l}\right) + \left(\vec{E}_4 \cdot \mathrm{d}\vec{l}\right) + \left(\vec{E}_4 \cdot \mathrm{d}\vec{l}\right) = -\frac{\partial \vec{B}_4}{\partial t} \cdot \mathrm{d}\vec{A}$$

Circulation 5:

$$-\left(\vec{E}_2 \cdot \mathrm{d}\vec{l}\right) + \left(\vec{E}_5 \cdot \mathrm{d}\vec{l}\right) - \left(\vec{E}_6 \cdot \mathrm{d}\vec{l}\right) + \left(\vec{E}_5 \cdot \mathrm{d}\vec{l}\right) - \left(\vec{E}_8 \cdot \mathrm{d}\vec{l}\right) + \left(\vec{E}_5 \cdot \mathrm{d}\vec{l}\right) - \left(\vec{E}_4 \cdot \mathrm{d}\vec{l}\right) + \left(\vec{E}_5 \cdot \mathrm{d}\vec{l}\right) = -\frac{\partial \vec{B}_5}{\partial t} \cdot \mathrm{d}\vec{A}$$

Circulation 6:

$$-\left(\vec{E}_3 \cdot \mathrm{d}\vec{l}\right) + \left(\vec{E}_6 \cdot \mathrm{d}\vec{l}\right) + \left(\vec{E}_6 \cdot \mathrm{d}\vec{l}\right) - \left(\vec{E}_9 \cdot \mathrm{d}\vec{l}\right) + \left(\vec{E}_6 \cdot \mathrm{d}\vec{l}\right) - \left(\vec{E}_5 \cdot \mathrm{d}\vec{l}\right) + \left(\vec{E}_6 \cdot \mathrm{d}\vec{l}\right) = -\frac{\partial \vec{B}_6}{\partial t} \cdot \mathrm{d}\vec{A}$$

Circulation 7:

$$-\left(\vec{E}_4 \cdot \mathrm{d}\vec{l}\right) + \left(\vec{E}_7 \cdot \mathrm{d}\vec{l}\right) - \left(\vec{E}_8 \cdot \mathrm{d}\vec{l}\right) + \left(\vec{E}_7 \cdot \mathrm{d}\vec{l}\right) + \left(\vec{E}_7 \cdot \mathrm{d}\vec{l}\right) + \left(\vec{E}_7 \cdot \mathrm{d}\vec{l}\right) = -\frac{\partial \vec{B}_7}{\partial t} \cdot \mathrm{d}\vec{A}$$

Circulation 8:

$$-\left(\vec{E}_5 \cdot \mathrm{d}\vec{l}\right) + \left(\vec{E}_8 \cdot \mathrm{d}\vec{l}\right) - \left(\vec{E}_9 \cdot \mathrm{d}\vec{l}\right) + \left(\vec{E}_8 \cdot \mathrm{d}\vec{l}\right) + \left(\vec{E}_8 \cdot \mathrm{d}\vec{l}\right) - \left(\vec{E}_7 \cdot \mathrm{d}\vec{l}\right) + \left(\vec{E}_8 \cdot \mathrm{d}\vec{l}\right) = -\frac{\partial \vec{B}_8}{\partial t} \cdot \mathrm{d}\vec{A}$$

Circulation 9:

$$-\left(\vec{E}_6 \cdot \mathrm{d}\vec{l}\right) + \left(\vec{E}_9 \cdot \mathrm{d}\vec{l}\right) + \left(\vec{E}_9 \cdot \mathrm{d}\vec{l}\right) + \left(\vec{E}_9 \cdot \mathrm{d}\vec{l}\right) - \left(\vec{E}_8 \cdot \mathrm{d}\vec{l}\right) + \left(\vec{E}_9 \cdot \mathrm{d}\vec{l}\right) = -\frac{\partial \vec{B}_9}{\partial t} \cdot \mathrm{d}\vec{A}$$

Adding these nine equations yields

$$+\left(\vec{E}_1 \cdot \mathrm{d}\vec{l}\right) - \left(\vec{E}_2 \cdot \mathrm{d}\vec{l}\right) + \left(\vec{E}_1 \cdot \mathrm{d}\vec{l}\right) - \left(\vec{E}_4 \cdot \mathrm{d}\vec{l}\right) + \left(\vec{E}_1 \cdot \mathrm{d}\vec{l}\right) + \left(\vec{E}_1 \cdot \mathrm{d}\vec{l}\right)$$

$$+\left(\vec{E}_2 \cdot \mathrm{d}\vec{l}\right) - \left(\vec{E}_3 \cdot \mathrm{d}\vec{l}\right) + \left(\vec{E}_2 \cdot \mathrm{d}\vec{l}\right) - \left(\vec{E}_5 \cdot \mathrm{d}\vec{l}\right) + \left(\vec{E}_2 \cdot \mathrm{d}\vec{l}\right) - \left(\vec{E}_1 \cdot \mathrm{d}\vec{l}\right) + \left(\vec{E}_2 \cdot \mathrm{d}\vec{l}\right)$$

$$+\left(\vec{E}_3 \cdot \mathrm{d}\vec{l}\right) + \left(\vec{E}_3 \cdot \mathrm{d}\vec{l}\right) - \left(\vec{E}_6 \cdot \mathrm{d}\vec{l}\right) + \left(\vec{E}_3 \cdot \mathrm{d}\vec{l}\right) - \left(\vec{E}_2 \cdot \mathrm{d}\vec{l}\right) + \left(\vec{E}_3 \cdot \mathrm{d}\vec{l}\right)$$

$$-\left(\vec{E}_1 \cdot \mathrm{d}\vec{l}\right) + \left(\vec{E}_4 \cdot \mathrm{d}\vec{l}\right) - \left(\vec{E}_5 \cdot \mathrm{d}\vec{l}\right) + \left(\vec{E}_4 \cdot \mathrm{d}\vec{l}\right) - \left(\vec{E}_7 \cdot \mathrm{d}\vec{l}\right) + \left(\vec{E}_4 \cdot \mathrm{d}\vec{l}\right) + \left(\vec{E}_4 \cdot \mathrm{d}\vec{l}\right)$$

$$-\left(\vec{E}_2 \cdot \mathrm{d}\vec{l}\right) + \left(\vec{E}_5 \cdot \mathrm{d}\vec{l}\right) - \left(\vec{E}_6 \cdot \mathrm{d}\vec{l}\right) + \left(\vec{E}_5 \cdot \mathrm{d}\vec{l}\right) - \left(\vec{E}_8 \cdot \mathrm{d}\vec{l}\right) + \left(\vec{E}_5 \cdot \mathrm{d}\vec{l}\right) - \left(\vec{E}_4 \cdot \mathrm{d}\vec{l}\right) + \left(\vec{E}_5 \cdot \mathrm{d}\vec{l}\right)$$

$$-\left(\vec{E}_3 \cdot d\vec{l}\right) + \left(\vec{E}_6 \cdot d\vec{l}\right) + \left(\vec{E}_6 \cdot d\vec{l}\right) - \left(\vec{E}_9 \cdot d\vec{l}\right) + \left(\vec{E}_6 \cdot d\vec{l}\right) - \left(\vec{E}_5 \cdot d\vec{l}\right) + \left(\vec{E}_6 \cdot d\vec{l}\right)$$

$$-\left(\vec{E}_4 \cdot d\vec{l}\right) + \left(\vec{E}_7 \cdot d\vec{l}\right) - \left(\vec{E}_8 \cdot d\vec{l}\right) + \left(\vec{E}_7 \cdot d\vec{l}\right) + \left(\vec{E}_7 \cdot d\vec{l}\right) + \left(\vec{E}_7 \cdot d\vec{l}\right)$$

$$-\left(\vec{E}_5 \cdot d\vec{l}\right) + \left(\vec{E}_8 \cdot d\vec{l}\right) - \left(\vec{E}_9 \cdot d\vec{l}\right) + \left(\vec{E}_8 \cdot d\vec{l}\right) + \left(\vec{E}_8 \cdot d\vec{l}\right) - \left(\vec{E}_7 \cdot d\vec{l}\right) + \left(\vec{E}_8 \cdot d\vec{l}\right)$$

$$-\left(\vec{E}_6 \cdot d\vec{l}\right) + \left(\vec{E}_9 \cdot d\vec{l}\right) + \left(\vec{E}_9 \cdot d\vec{l}\right) + \left(\vec{E}_9 \cdot d\vec{l}\right) - \left(\vec{E}_8 \cdot d\vec{l}\right) + \left(\vec{E}_9 \cdot d\vec{l}\right)$$

$$= -\frac{\partial \vec{B}_1}{\partial t} \cdot d\vec{A} - \frac{\partial \vec{B}_2}{\partial t} \cdot d\vec{A} - \frac{\partial \vec{B}_3}{\partial t} \cdot d\vec{A} - \frac{\partial \vec{B}_4}{\partial t} \cdot d\vec{A} - \frac{\partial \vec{B}_5}{\partial t} \cdot d\vec{A} - \frac{\partial \vec{B}_6}{\partial t} \cdot d\vec{A} - \frac{\partial \vec{B}_7}{\partial t} \cdot d\vec{A} - \frac{\partial \vec{B}_8}{\partial t} \cdot d\vec{A} - \frac{\partial \vec{B}_9}{\partial t} \cdot d\vec{A}$$

and

$$\left(\vec{E}_1 \cdot d\vec{l}\right) + \left(\vec{E}_2 \cdot d\vec{l}\right) + \left(\vec{E}_3 \cdot d\vec{l}\right) + \left(\vec{E}_3 \cdot d\vec{l}\right) + \left(\vec{E}_6 \cdot d\vec{l}\right) + \left(\vec{E}_9 \cdot d\vec{l}\right) + \left(\vec{E}_9 \cdot d\vec{l}\right)$$

$$+\left(\vec{E}_8 \cdot d\vec{l}\right) + \left(\vec{E}_7 \cdot d\vec{l}\right) + \left(\vec{E}_7 \cdot d\vec{l}\right) + \left(\vec{E}_4 \cdot d\vec{l}\right) + \left(\vec{E}_1 \cdot d\vec{l}\right) =$$

$$-\frac{\partial \vec{B}_1}{\partial t} \cdot d\vec{A} - \frac{\partial \vec{B}_2}{\partial t} \cdot d\vec{A} - \frac{\partial \vec{B}_3}{\partial t} \cdot d\vec{A} - \frac{\partial \vec{B}_4}{\partial t} \cdot d\vec{A} - \frac{\partial \vec{B}_5}{\partial t} \cdot d\vec{A}$$

$$-\frac{\partial \vec{B}_6}{\partial t} \cdot d\vec{A} - \frac{\partial \vec{B}_7}{\partial t} \cdot d\vec{A} - \frac{\partial \vec{B}_8}{\partial t} \cdot d\vec{A} - \frac{\partial \vec{B}_9}{\partial t} \cdot d\vec{A}$$

$$(4.18)$$

The left side of Eq. (4.18) is the path along the outer contour of all nine infinitesimal squares with the side length $d\vec{l}$ multiplied by the electric field $\vec{E}$ of each section. The contour of the entire surface is shown as a solid bold line in Fig. 4.8. Starting point is the upper left corner. The right side contains the sum of the temporal changes of the magnetic flux in the nine partial surfaces.

The general equation for the derived fact is

$$\oint_C \vec{E} \cdot d\vec{l} = -\iint_A \frac{\partial \vec{B}}{\partial t} \cdot d\vec{A} \qquad (4.19)$$

Equation (4.19) is the law of induction in its integral notation. With the theorem of Stokes (cf. 3.93), we have

$$\oint_C \vec{E} \cdot d\vec{l} = \iint_A \nabla \times \vec{E} \cdot d\vec{A} = -\iint_A \frac{\partial \vec{B}}{\partial t} \cdot d\vec{A} \qquad (4.20)$$

and thus

$$\nabla \times \vec{E} = -\frac{\partial \vec{B}}{\partial t} \qquad (4.21)$$

Fig. 4.9 $\vec{E}$ and $d\vec{B}/dt$ of the
second Maxwell's equation [6]

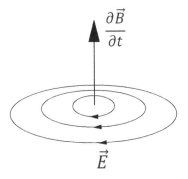

This is Maxwell's second equation. It can be understood as the differential form of the
law of induction. Eq. (4.21) says:

> Where there is a time-varying magnetic flux density in space, there is also an electric field.

In vector-analytical formulation, Eq. (4.18) contains the following statement.

> The electric field has vortices, or curls, at locations where the magnetic field changes in time. The magnitude of the curl of the electric field depends on the rate of change of the magnetic field.

Figure 4.9 illustrates this fact. The electric field lines and the vector of the temporal
change of the magnetic flux density form a left-hand screw.

To conclude this section, we look at Eq. (4.16) in more detail. First, we return to
Eq. (3.16)

$$\vec{E} = -\left(\vec{v} \times \vec{B}\right)$$

According to (3.16), the electric field $\vec{E}$ is the result of the movement of a conductor
with the velocity $\vec{v}$ in a magnetic field with the flux density $\vec{B}$. According to the previous considerations, the presence of a conductor is not required for the generation of an
electric field. Even without an electrical conductor and without the presence of electrical
charges, an electric field is generated along a contour C, if this contour moves in a magnetic field.

When deriving the law of induction (4.19), no prerequisite was made regarding the cause of the temporal change of the magnetic flux density within contour C.

Equation (4.19) includes the occurrence of an electric field along contour C, both by a temporal change of the magnetic flux density within contour C without the contour moving, and also as a result of the movement of the contour in the magnetic field. In order to formally separate the occurrence of an electric field by a change of the flux from the occurrence by a movement of the contour, Eq. (4.16) must be modified.

The part which generates an electric field due to a temporal change in the magnetic flux density within contour C without the contour moving in the magnetic field is determined by

$$\oint_C \vec{E} \cdot d\vec{l} = -\frac{d}{dt} \iint_A \vec{B} \cdot d\vec{A} \tag{4.22}$$

In order to include the occurrence of the electric field along the contour C due to the movement of this contour in the magnetic field, the integral at the right side of this equation must be supplemented according to (3.16) by the term

$$-\int_C \left(\vec{v} \times \vec{B} \right) \cdot d\vec{l}$$

Now we have

$$\oint_C \vec{E} \cdot d\vec{l} = -\frac{d}{dt} \iint_A \vec{B} \cdot d\vec{A} - \int_C \left(\vec{v} \times \vec{B} \right) \cdot d\vec{l}$$

and

$$\oint_C \left[\vec{E} + \left(\vec{v} \times \vec{B} \right) \right] \cdot d\vec{l} = -\frac{d}{dt} \iint_A \vec{B} \cdot d\vec{A} \tag{4.23}$$

The component $\left(\vec{v} \times \vec{B} \right)$ of the left integral of this equation represents the electric field due to the movement of the contour element of contour C within the magnetic field of flux density $\vec{B}$. Since all elements of the contour may move at different speeds, this term also includes the generation of an electric field due to a change in the shape of the contour.

Equation (4.23), like Eq. (4.19), is the correct notation of Faraday's law of induction. Methods of vector analysis can be used to demonstrate that Eq. (4.19) can be converted into Eq. (4.23)[8].

---

[8]Flanders, H.: Differential under the integral sign, American Mathematical Monthly (6), pp. 615–627.

4.4   Continuity Equation

The continuity equation formulates the following statement in mathematical form:

The electrical charge in a volume can only change if a charge either flows out of the volume across the enveloping surface of the volume or flows into the volume. Overall, the electrical charge remains unchanged.

Figure 4.10 shows a charge Q in a volume V. The charge ΔQ_1 enters the volume, and the charge ΔQ_2 leaves the volume. If $\Delta Q_1 \neq \Delta Q_2$ the quantity of charge inside the volume changes. According to (1.28), the transport of charge is equivalent to a current flow. A reduction of the charge of ∂Q within a time period ∂t corresponds to a current I, which exits from the enveloping surface of the volume of

$$I = -\frac{\partial Q(\vec{r}, t)}{\partial t} \tag{4.24}$$

The negative sign in Eq. (4.24) indicates that the normal vector of the enveloping surface of the volume is oriented outward. As a result, a decrease in charge within the volume V results in an outwardly directed and thus positive current. Equation (4.24) is the "continuity equation" ([3], p. 66).

In Fig. 4.10, no assumption was made with regard to the charge distribution in volume V. The volume may contain arbitrarily distributed charge. The same applies to the type of

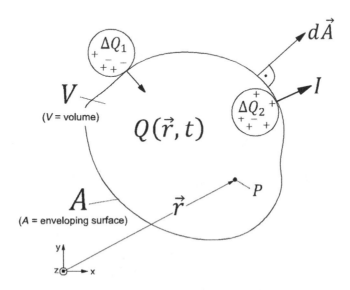

Fig. 4.10 Conservation of charge

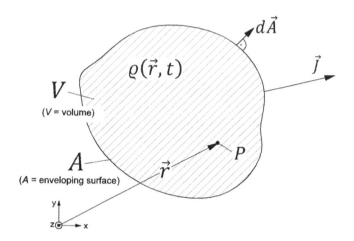

Fig. 4.11 Conservation of charge with continuous charge distribution

current flow. If a spatial, possibly continuous distribution of the charge, is assumed in the volume, Eq. (4.24) has to be adapted accordingly. The charge $Q(\vec{r}, t)$ is to be replaced by the space charge density $\varrho\ (\vec{r}, t)$, and the current I is to be replaced by the current density $\vec{J}$ (see Fig. 4.11).

The total current exiting from the volume V through the closed enveloping surface A to the outside is

$$I = \oiint_A \vec{J} \cdot \mathrm{d}\vec{A} \tag{4.25}$$

The temporal decrease of charge within the volume V results in a current I out of the enveloping surface

$$I = -\frac{\partial Q}{\partial t} = -\frac{\partial}{\partial t} \iiint_V \varrho \cdot \mathrm{d}V = -\iiint_V \frac{\partial \varrho}{\partial t} \cdot \mathrm{d}V \tag{4.26}$$

By applying the theorem of Gauss (2.35), we have

$$\iiint_V \frac{\partial \varrho}{\partial t} \cdot \mathrm{d}V = -\oiint_A \vec{J} \cdot \mathrm{d}\vec{A} = \iiint_V -\nabla \cdot \vec{J} \cdot \mathrm{d}V \tag{4.27}$$

As the integrands of the volume integrals in this equation are equal,

$$\nabla \cdot \vec{J} = -\frac{\partial \varrho}{\partial t} \tag{4.28}$$

Equation (4.28) is the continuity equation in its differential form. This means the following:

A temporal change (decrease or increase) of the charge density in the elementary volume is equivalent to the "divergence" of the current density, i.e., to the rate of change (decrease or increase) of the current density (see comment on Eq. (2.29)).

Figure 4.12 shows a charge cloud moving through the enveloping surface A into the volume V. According to (2.33), the relationship between charge density ϱ and electrical flux density $\vec{D}$ within the volume V is

$$\iiint_V \varrho \cdot \mathrm{d}V = \iiint_V \left(\nabla \cdot \vec{D}\right) \cdot \mathrm{d}V \tag{4.29}$$

The volume V is arbitrary, thus

$$\nabla \cdot \vec{D} = \varrho \tag{4.30}$$

Equation (4.30) corresponds to (2.29). For the moving charge cloud, Eq. (4.30) can be applied for any time t. By differentiating (4.30) with respect to time, it becomes clear how the temporal change of the electrical flux density affects the change of the charge density

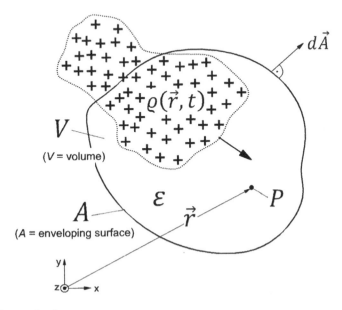

Fig. 4.12 Charge cloud

$$\nabla \cdot \left(\frac{\partial \vec{D}}{\partial t} \right) = \frac{\partial \varrho}{\partial t} \tag{4.31}$$

or

$$\nabla \cdot \left(\frac{\partial \vec{D}}{\partial t} \right) - \frac{\partial \varrho}{\partial t} = 0 \tag{4.32}$$

Putting (4.28) into this equation gives another version of the continuity equation

$$\nabla \cdot \left(\frac{\partial \vec{D}}{\partial t} + \vec{J} \right) = 0 \tag{4.33}$$

The temporal change of the electrical flux density $d\vec{D}/dt$ can be interpreted as a current density. According to Maxwell, it is referred to as displacement current density. This term is explained in more detail in the following section.

4.5   Displacement Current Density

Figure 4.13 shows an open DC circuit with a capacitor. In this figure, the two electrodes of the capacitor, and their distance from each other, are greatly enlarged for a better representation of the current density. The circuit corresponds to the circuit in Fig. 2.8. If the switch S is closed at time $t = 0$, the capacitor C will charge. With the circuit data in Fig. 2.8, the charge current $i(t)$ drops after closing switch S from 6 mA to almost 0 mA within about 60 μs (see diagram in Fig. 2.9). The current $i(t)$ results in a current density $\vec{J}$ in the electrodes of the capacitor. During the charging process, charge is supplied to

Fig. 4.13 Open circuit with
capacitor

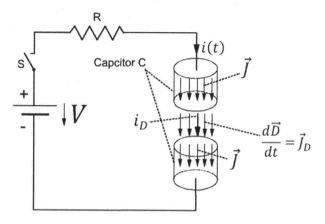

the positive electrode, while charge flows off from the negative electrode. As a result, the electrical flux density in the field between the electrodes of the capacitor also changes as long as a current flows.

The temporal change of the flux density $\partial\vec{D}/\partial t$ according to (4.33) is a current density, the so-called displacement current density

$$\frac{\partial\vec{D}}{\partial t} = \vec{J}_D$$

The displacement current density is present in the electrical field between the capacitor plates as long as the electrical flux density changes with time, i.e., as long as a current $i(t)$ flows in the connecting wires of the capacitor, and the current changes with time. The current $i(t)$ "continues to flow" in the electric field between the capacitor plates as a displacement current i_D with the displacement current density $\vec{J}_D$. As a result, the current entering a volume containing parts of the circuit is equal to the current leaving the volume. This satisfies the continuity equation for the circuit shown in Fig. 4.13.

The same applies to the AC circuit shown in Fig. 4.14. The capacitor is repeatedly charged and discharged, so that a displacement current exists always in the electric field between the electrodes of the capacitor. In this case, the current density $\vec{J}$ in the conductors continues as displacement current density $\partial\vec{D}/\partial t = \vec{J}_D$ in the electric field between the capacitor electrodes, so that the originally open circuit is closed by the displacement current i_D and the condition

$$\nabla \cdot \left(\frac{\partial\vec{D}}{\partial t} + \vec{J} \right) = 0$$

of the continuity Eq. (4.33) is satisfied.

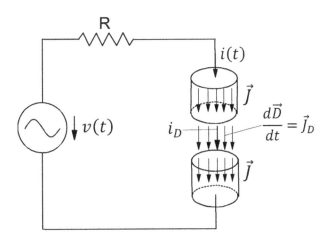

Fig. 4.14 AC circuit with capacitor

4.6   First Maxwell's Equation

The following relationship applies to the electric current I_{con} which is concatenated to the surface A, i.e., flows through this surface

$$I_{\mathrm{con}} = \iint_A \vec{J} \cdot \vec{n}_A \cdot \mathrm{d}A$$

With this relation, Ampère's law takes the following form according to (3.45):

$$\oint_C \vec{H} \cdot \mathrm{d}\vec{s} = I_{\mathrm{con}} = I = \iint_A \vec{J} \cdot \vec{n}_A \cdot \mathrm{d}A \qquad (4.34)$$

This equation contains the following statement:

> The integral of the magnetic field $\vec{H}$ along a closed contour C is equal to the current I_{con} flowing through the surface A formed by this contour.

Figure 4.15 shows an AC circuit with a capacitor. Ampère's law should also be valid for this circuit. The two bell-shaped surfaces A_1 and A_2 have the same contour C. The integral over the surface A_1 with the contour C is obviously different from the integral over the surface A_2 with the same contour C. Through the surface, A_1 flows the electric current $i(t)$. Ampère's law is therefore fulfilled

$$\oint_{C/A_1} \vec{H} \cdot \mathrm{d}\vec{s} = i(t)$$

Fig. 4.15 On Ampère's law

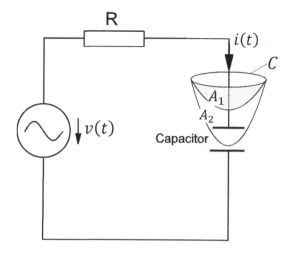

Obviously, no current flows through the surface A_2. The integral over the contour C of this surface should therefore be equal to zero

$$\oint_{C/A_2} \vec{H} \cdot d\vec{s} = 0$$

Since both surfaces have the same contour C, both equations contradict each other. Maxwell has solved this contradiction by postulating that the displacement current $i_D(t)$ or the displacement current density $\partial \vec{D}/\partial t$ between the capacitor electrodes, as well as the conductor current $i(t)$, is linked to a magnetic field. For contour C with A_2

$$\oint_{C/A_2} \vec{H} \cdot d\vec{s} = \iint_{A_2} \frac{\partial \vec{D}}{\partial t} \cdot \vec{n}_A \cdot dA$$

Ampère's law according to Eq. (4.34) must therefore be supplemented by the displacement current density for time-varying electric fields

$$\oint_{C} \vec{H} \cdot d\vec{s} = \iint_{A} \left(\vec{J} + \frac{\partial \vec{D}}{\partial t} \right) \cdot \vec{n}_A \cdot dA \qquad (4.35)$$

A vector-analytical approach leads to the same result. According to (3.60), the differential form of Ampère's law for stationary magnetic fields is

$$\nabla \times \vec{H} = \vec{J} \qquad (4.36)$$

Ampère's law cannot be valid generally in this form in the sense of vector analysis. According to Eq. (3.87), we have

$$\nabla \cdot \left(\nabla \times \vec{H} \right) = 0 \qquad (4.37)$$

And according to Eq. (3.60)

$$\nabla \times \vec{H} = \vec{J}$$

Therefore, (4.37) is only valid if

$$\nabla \cdot \vec{J} = 0$$

i.e., only if the current density is source free. However, this is not true because of the continuity Eq. (4.28)

$$\nabla \cdot \vec{J} = -\frac{\partial \varrho}{\partial t}$$

Only by including the existence of the displacement current, according to Eq. (4.33), the requirement for source freedom can be fulfilled

$$\nabla \cdot \left(\frac{\partial \vec{D}}{\partial t} + \vec{J} \right) = 0$$

Even from a vector-analytical point of view, the current density for time-varying fields must therefore be supplemented by the partial derivate of the electrical flux density with respect to time, i.e., by the displacement current density.

The following equation replaces Ampère's law of (3.59 and 3.60), respectively, in the case of time-varying fields

$$\nabla \times \vec{H} = \frac{\partial \vec{D}}{\partial t} + \vec{J} \qquad\qquad (4.38)$$

This equation satisfies the condition (4.37)

$$\nabla \cdot \left(\nabla \times \vec{H} \right) = 0$$

For a static field $\partial \vec{D} / \partial t = 0$. Equation (4.38) corresponds to Ampère's law in Eq. (3.59).

Equation (4.38) is called the first Maxwell's equation. It formulates the physical fact that both the current density and the temporally variable electrical flux density are linked to, or accompanied by, a magnetic field. Figure 4.16 illustrates this fact. The right part of the figure refers to a non-conductive medium, e.g., vacuum. In this case, $\vec{J} = 0$ and a time-varying electric field

$$\varepsilon \frac{\partial \vec{E}}{\partial t} = \frac{\partial \vec{D}}{\partial t}$$

is linked to a magnetic field.

Another way to illustrate the statement of Maxwell's first equation is the circuit arrangement shown in Fig. 4.17. An alternating current flows in the circuit. Both the magnetic field generated by the current i(t) and the magnetic field generated by the displacement current density $\partial \vec{D} / \partial t$, could in principle be measured with a Rogowski inductor as a magnetic voltmeter.

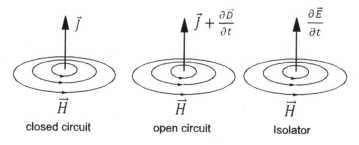

Fig. 4.16 Interpretation of the first Maxwell's equation [6]

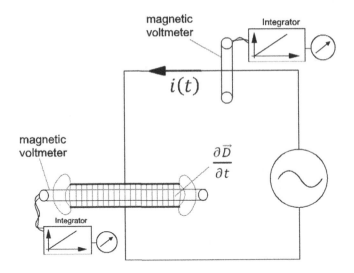

Fig. 4.17 Circuit to explain the first Maxwell's equation

Fig. 4.18 Interpretation of the first and second Maxwell's equations for vacuum [6]

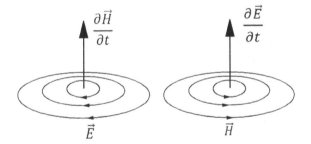

To conclude this section, Fig. 4.18 compares the statements of the first and second Maxwell's equations in the non-conductive medium ($\vec{J} = 0, \vec{E} = \vec{D}/\varepsilon$ and $\vec{H} = \vec{B}/\mu$). The following symmetry can be seen in both figures: According to Maxwell's first equation, the time-varying magnetic field is linked to or accompanied by an electric field, and according to Maxwell's second equation, a time-varying electric field is linked to a magnetic field. This relationship is, as will be shown, the condition for the propagation of electromagnetic waves in a non-conducting medium, e.g., in a vacuum. Obviously, a magnetic field can be generated without the need for a current-carrying conductor at this location.

4.6.1 Summary of the Equations

At this point, it seems useful to summarize the equations developed in the previous sections and chapters.

Field Equations
First Maxwell's Eq. (4.38)

$$\nabla \times \underline{\vec{H}} = \frac{\partial \vec{D}}{\partial t} + \vec{J} \tag{4.39}$$

Second Maxwell's equation or Faraday's law of induction (4.21)

$$\nabla \times \vec{E} = -\frac{\partial \vec{B}}{\partial t} \tag{4.40}$$

Continuity equations Equation (4.33):

$$\nabla \cdot \left(\frac{\partial \vec{D}}{\partial t} + \vec{J} \right) = 0 \tag{4.41}$$

Equation (2.29):

$$\nabla \cdot \vec{D} = \varrho \tag{4.42}$$

Equation (3.8):

$$\nabla \cdot \vec{B} = 0 \tag{4.43}$$

Equations for static, linear, and isotropic media
Equation (2.11):

$$\vec{D} = \varepsilon \cdot \vec{E} \tag{4.44}$$

Equation (1.22):

$$\vec{J} = \sigma \cdot \vec{E} \tag{4.45}$$

Equation (3.46):

$$\vec{B} = \mu \cdot \vec{H} \tag{4.46}$$

4.7   Time-Harmonic Fields

In the next chapters, we will assume that all electromagnetic fields are located within linear media. As a result, time-varying fields can be represented by Fourier series or by Fourier integrals. This means that electric and magnetic fields can be described by sin and cos functions; furthermore, it allows us to use complex phasor arithmetic.

The following considerations can be limited to temporally cosinusoidal (or sinusodial) changes of the fields. In accordance with the phasor arithmetic, it is useful to use the complex arithmetic. For example, the electric field vector $\vec{E}(\vec{r}, t)$ is

$$\vec{E}(\vec{r}, t) = \vec{E}(\vec{r}) \cdot \cos(\omega \cdot t + \varphi_0) \tag{4.47}$$

In Eq. (4.47), $\vec{r}$ is the space vector of the electric field and φ_0 the phase angle of the cosine time dependence with the angular frequency ω. The complex representation of Eq. (4.47) is[9]

$$\vec{E}(\vec{r}, t) = Re\left\{\vec{\underline{E}}(\vec{r}) \cdot e^{j \cdot (\omega \cdot t + \varphi_0)}\right\}$$

$$\vec{E}(\vec{r}, t) = Re\left\{\vec{E}(\vec{r}) \cdot e^{j \cdot \varphi_0} \cdot e^{j \cdot \omega \cdot t}\right\} = Re\left\{\vec{\underline{E}}(\vec{r}) \cdot e^{j \cdot \omega \cdot t}\right\} \tag{4.48}$$

Also

$$Re\left\{\vec{\underline{E}}(\vec{r}) \cdot e^{j \cdot \omega \cdot t}\right\} = Re\left\{\vec{E}(\vec{r}) \cdot \left[\cos\left(\omega \cdot t + \varphi_0\right) + j \cdot \sin\left(\omega \cdot t + \varphi_0\right)\right]\right\}$$

In Eq. (4.48), $\vec{\underline{E}}(\vec{r})$ is the complex field vector or phasor of the electric field. It includes the phase angle φ_0. To simplify the calculation, the complex field vector $\vec{\underline{E}}(\vec{r})$ is used for symbolic calculation, i.e., the factor $e^{j \cdot \omega \cdot t}$ is suppressed in the calculation. This is permissible because the vector operations only contain derivatives with respect to the space coordinates. However, when differentiating with respect to time, it must be taken into account that the factor $e^{j \cdot \omega \cdot t}$ was omitted.

The first and second derivatives of the electric field with respect to time t are

$$\frac{\partial\left(\vec{\underline{E}}(\vec{r}) \cdot e^{j \cdot \omega \cdot t}\right)}{\partial t} = j \cdot \omega \cdot \vec{\underline{E}}(\vec{r}) \cdot e^{j \cdot \omega \cdot t} \tag{4.49}$$

and

$$\frac{\partial^2\left(\vec{\underline{E}}(\vec{r}) \cdot e^{j \cdot \omega \cdot t}\right)}{\partial t^2} = -\omega^2 \cdot \vec{\underline{E}}(\vec{r}) \cdot e^{j \cdot \omega \cdot t} \tag{4.50}$$

respectively.

The first derivative with respect to t therefore corresponds to a multiplication with $(j \cdot \omega)$ and the second derivative to a multiplication with $(-\omega^2)$. The same considerations apply to the magnetic field vector.

For time-harmonic dependence, Maxwell's equations (4.39) to (4.42) can be written as

Field equations

$$\nabla \times \vec{\underline{H}} = \vec{\underline{J}} + j \cdot \omega \cdot \vec{\underline{D}} \tag{4.51}$$

$$\nabla \times \vec{\underline{E}} = -j \cdot \omega \cdot \vec{\underline{B}} \tag{4.52}$$

---

[9]$Re\left\{\vec{E}(\vec{r}) \cdot e^{j \cdot (\omega \cdot t + \varphi_0)}\right\}$ means: real component of $\vec{E}(\vec{r}) \cdot e^{j \cdot (\omega \cdot t + \varphi_0)}$.

Continuity equations

$$\nabla \cdot \left(j \cdot \omega \cdot \vec{\underline{D}} + \vec{\underline{J}} \right) = 0 \tag{4.53}$$

$$\nabla \cdot \vec{\underline{D}} = \varrho \text{ or } \nabla \cdot \vec{\underline{E}} = \frac{\varrho}{\varepsilon} \tag{4.54}$$

The time-harmonic dependence can be restored explicitly, by multiplying the result of the calculation according to Eq. (4.48) with $e^{j \cdot \omega \cdot t}$ and then taking the real part.

4.8 Wave Equations

A wave is a spatially propagating change of a space- and time-dependent physical quantity. In electrodynamics, these are the electric and the magnetic fields. In the following sections, electromagnetic waves are considered propagating in a vacuum. These considerations begin with the time function for the one-dimensional wave propagation, independent of the propagation medium. For the calculation of the electromagnetic waves, we have Maxwell's equations at our disposal. These are partial differential equations in the three-dimensional space. Nevertheless, the first step is to derive the differential equation for one-dimensional fields.

Starting point for the wave equations is the function

$$f\left(t - \frac{z}{c} \right) = V \cdot e^{-a \cdot z} \cdot \cos \left[\left(t - \frac{z}{c} \right) + \varphi \right] \tag{4.55}$$

The factor $\left(V \cdot e^{-a \cdot z} \right)$ is the amplitude of a voltage wave that decreases as it progresses in z-direction. The graph of this function is shown in Fig. 4.19 for the two times t_1 and t_2. The marked maximum at time t_1 is z_1 and at time t_2 is z_2. It can be seen that the function $f\left(t - \frac{z}{c} \right)$ is a one-dimensional wave propagating, with velocity c, in positive z-direction. The path difference $(z_2 - z_1)$ is

$$(z_2 - z_1) = (t_2 - t_1) \cdot c \tag{4.56}$$

Therefore, the velocity of the propagation is

$$c = \frac{(z_2 - z_1)}{(t_2 - t_1)} \tag{4.57}$$

Figure 4.19 shows the propagation of a wave with decreasing magnitudes, while z is increasing.

For a wave of the same type, as shown in Fig. 4.19, but traveling into the negative z-direction, we have

$$g\left(t + \frac{z}{c} \right) = V \cdot e^{-a \cdot x} \cdot \cos \left[\left(t + \frac{z}{c} \right) + \varphi \right] \tag{4.58}$$

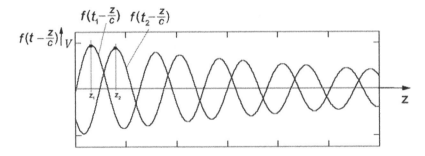

Fig. 4.19 Wave propagation

By superimposing a wave propagating into the positive z-direction (Eq. 4.55) and a wave propagating into the negative z-direction (Eq. 4.58), different waveforms including standing waves can be generated

$$w(z,t) = f\left(t - \frac{z}{c}\right) + g\left(t + \frac{z}{c}\right) \tag{4.59}$$

With

$$u = \left(t - \frac{z}{c}\right) \tag{4.60}$$

and

$$v = \left(t + \frac{z}{c}\right) \tag{4.61}$$

we get

$$\frac{\partial w}{\partial z} = \frac{df}{du} \cdot \frac{\partial u}{\partial z} + \frac{dg}{dv} \cdot \frac{\partial v}{\partial z} \tag{4.62}$$

From Eq. (4.59) with Eqs. (4.60) and (4.61) follows

$$\frac{\partial w}{\partial z} = \frac{df}{du} \cdot \left(-\frac{1}{c}\right) + \frac{dg}{dv} \cdot \left(\frac{1}{c}\right)$$

respectively

$$\frac{\partial w}{\partial z} = -\frac{1}{c} \cdot \frac{df}{du} + \frac{1}{c} \cdot \frac{dg}{dv} \tag{4.63}$$

The partial derivative of this differential equation with respect to z is

$$\frac{\partial^2 w}{\partial z^2} = -\frac{1}{c} \cdot \frac{d^2 f}{du^2} \cdot \frac{\partial u}{\partial z} + \frac{1}{c} \cdot \frac{d^2 g}{dv^2} \cdot \frac{\partial v}{\partial z}$$

Given (4.59, 4.60, and 4.61), we have

$$\frac{\partial^2 w}{\partial z^2} = -\frac{1}{c} \cdot \frac{d^2 f}{du^2} \cdot \left(-\frac{1}{c}\right) + \frac{1}{c} \cdot \frac{d^2 g}{dv^2} \cdot \left(\frac{1}{c}\right)$$

or

$$\frac{\partial^2 w}{\partial z^2} = \frac{1}{c^2} \cdot \left(\frac{d^2 f}{du^2} + \frac{d^2 g}{dv^2}\right) \tag{4.64}$$

The second derivative of the function $w(z, t)$ with respect to t is

$$\frac{\partial^2 w}{\partial t^2} = \frac{d^2 f}{du^2} + \frac{d^2 g}{dv^2} \tag{4.65}$$

With (4.64), the following differential equation is obtained:

$$\frac{\partial^2 w}{\partial z^2} - \frac{1}{c^2} \cdot \frac{\partial^2 w}{\partial t^2} = 0 \tag{4.66}$$

Equation (4.66) is the so-called wave equation for the propagation of a one-dimensional wave. It is the simplest form of the wave equation. Equation (4.59) is the general solution of this partial differential equation.

In the case that the functions $f(z, t)$ and $g(z, t)$ are time-harmonic functions according to Eq. (4.58), also function $w(z, t)$ is a time-harmonic function, and thus (4.66) changes into

$$\frac{\partial^2 \underline{w}}{\partial z^2} + \frac{\omega^2}{c^2} \cdot \underline{w} = 0 \tag{4.67}$$

Equations (4.66 and 4.67) are wave equations of one-dimensional scalar fields. Figure 4.20 shows the wave of a two-dimensional scalar field. A water wave, for example, is a wave of this kind. The scalar is the height of the water, which depends on distance to the location of excitation of the wave and on time.

Waves of a three-dimensional field cannot be represented as a two-dimensional image. For the image of the wave of a three-dimensional scalar field, we have to imagine a transformation from Fig. 4.20 to three-dimensional space.

Fig. 4.20 Image of the wave of a two-dimensional scalar field (*source*: Fotolia_41270407_L)

Equation (4.67) can be used to derive the wave equation of a three-dimensional scalar field for time-harmonic dependence. Function $\underline{w}(z)$ is replaced by function $\underline{w}(x, y, z)$, and the partial derivative in z-direction is replaced by the spatial derivative, i.e., the gradient (grad $\underline{w} = \nabla \underline{w}$). The result of the spatial derivative, i.e., the result of the gradient operation, is a vector. Since the wave equation is a second-order partial differential equation, the gradient operation is once more differentiated. The spatial derivative of the vector $\nabla(\underline{w}(x, y, z))$ is the divergence of this vector, i.e., in the case of Cartesian coordinates, the change of the x-coordinate of the vector $\nabla(\underline{w}(x, y, z))$ into x-direction plus the change of the y-coordinate into y-direction plus the change of the z-coordinate into z-direction. This is the second spatial derivative of the function $\underline{w}(x, y, z)$. The result is a scalar.

Thus, the wave equation of a time-harmonic wave of a three-dimensional scalar field is

$$\nabla \cdot (\nabla(\underline{w}(x, y, z))) + \frac{\omega^2}{c^2} \cdot \underline{w}(x, y, z) = 0 \qquad (4.68)$$

For the operation $\nabla \cdot (\nabla \underline{w})$ in this equation (see Eqs. 1.7 and 2.28), we have

$$\nabla \cdot (\nabla \underline{w}) = \frac{\partial}{\partial x}\left(\frac{\partial \underline{w}}{\partial x}\right) + \frac{\partial}{\partial y}\left(\frac{\partial \underline{w}}{\partial y}\right) + \frac{\partial}{\partial z}\left(\frac{\partial \underline{w}}{\partial z}\right)$$

The operation

$$\nabla \cdot (\nabla \underline{w})$$

can be written as

$$\nabla \cdot (\nabla \underline{w}) = \nabla^2 \underline{w} \qquad (4.69)$$

Thus, equation (4.68) changes into

$$\nabla^2 \underline{w} + \frac{\omega^2}{c^2} \cdot \underline{w} = 0 \qquad (4.70)$$

An electromagnetic wave is a spatially propagating change of the space- and time-dependent magnetic field vector and the space- and time-dependent electric field vector. Therefore, for three-dimensional vector fields, equation (4.67) must be modified as

$$\nabla^2 \vec{\underline{w}} + \frac{\omega^2}{c^2} \cdot \vec{\underline{w}} = 0 \qquad (4.71)$$

Application of the operator ∇^2 on a vector field is defined in Cartesian coordinates as

$$\nabla^2 \vec{\underline{w}} = \left(\nabla^2 \underline{w}_x\right) \cdot \vec{e}_x + \left(\nabla^2 \underline{w}_y\right) \cdot \vec{e}_y + \left(\nabla^2 \underline{w}_z\right) \cdot \vec{e}_z \qquad (4.72)$$

The homogeneous partial differential Eqs. (4.70 and 4.71) take no account of the way the fields are excited. They are valid for source-free space. The type of wave propagation is only determined by the boundary conditions. With waveguides such as lossless waveguides, the tangential components of the electric field have to disappear on the ideally conducting waveguide surfaces and thus the propagation only occurs in the direction of the waveguide axis.

In contrast to the homogeneous partial differential Eqs. (4.70 and 4.71), the following Eq. (4.73) contains a so-called excitation vector $\underline{\vec{s}}$. It determines the properties of the transmitter that is responsible for the excitation of the electromagnetic wave.

$$\nabla^2 \underline{\vec{w}} + \frac{\omega^2}{c^2} \cdot \underline{\vec{w}} = \underline{\vec{s}} \tag{4.73}$$

Equation (4.73) is equivalent to three scalar equations, where all components are dependent on the three coordinates x, y, and z

$$\nabla^2 \underline{w}_x + \frac{\omega^2}{c^2} \cdot \underline{w}_x = \underline{s}_x$$

$$\nabla^2 \underline{w}_y + \frac{\omega^2}{c^2} \cdot \underline{w}_y = \underline{s}_y \tag{4.74}$$

$$\nabla^2 \underline{w}_z + \frac{\omega^2}{c^2} \cdot \underline{w}_z = \underline{s}_z$$

4.9 Inhomogeneous Wave Equation for the Electric and Magnetic Field Vectors

In this section, it is shown that in Maxwell's equations the propagation of electromagnetic waves is implemented, which is not easily to be seen in this system of equations. Let us start with Maxwell's first Eq. (4.38)

$$\nabla \times \vec{H} = \frac{\partial \vec{D}}{\partial t} + \vec{J}$$

respectively

$$\nabla \times \vec{B} = \varepsilon \cdot \mu \cdot \frac{\partial \vec{E}}{\partial t} + \mu \cdot \vec{J}$$

For the case of free space (vacuum: $\mu = \mu_0$ and $\varepsilon = \varepsilon_0$)

$$\nabla \times \vec{B} = \varepsilon_0 \cdot \mu_0 \cdot \frac{\partial \vec{E}}{\partial t} + \mu_0 \cdot \vec{J} \tag{4.75}$$

From the second Maxwell's Eq. (4.21)

$$\nabla \times \vec{E} = -\frac{\partial \vec{B}}{\partial t}$$

with another curl

$$\nabla \times \left(\nabla \times \vec{E}\right) = -\frac{\partial}{\partial t}\left(\nabla \times \vec{B}\right)$$

In this equation, we insert $\nabla \times \vec{B}$ from Eq. (4.75)

$$\nabla \times \left(\nabla \times \vec{E}\right) = -\varepsilon_0 \cdot \mu_0 \cdot \frac{\partial^2 \vec{E}}{\partial t^2} - \mu_0 \cdot \frac{\partial \vec{J}}{\partial t} \tag{4.76}$$

$$\nabla \times \left(\nabla \times \vec{E}\right) = -\frac{\partial}{\partial t}\left(\varepsilon_0 \cdot \mu_0 \cdot \frac{\partial \vec{E}}{\partial t} + \mu_0 \cdot \vec{J}\right)$$

With a relationship from the vector analysis rules (see 3.88)

$$\nabla \times \left(\nabla \times \vec{E}\right) = \nabla\left(\nabla \cdot \vec{E}\right) - \nabla^2 \vec{E} \tag{4.77}$$

and Eqs. (4.76 and 4.77)

$$-\varepsilon_0 \cdot \mu_0 \cdot \frac{\partial^2 \vec{E}}{\partial t^2} - \mu_0 \cdot \frac{\partial \vec{J}}{\partial t} = \nabla\left(\nabla \cdot \vec{E}\right) - \nabla^2 \vec{E}$$

Now we insert (4.42) into this equation

$$\nabla \cdot \vec{D} = \varrho$$

or

$$\nabla \cdot \vec{E} = \frac{\varrho}{\varepsilon_0}$$

yielding

$$-\varepsilon_0 \cdot \mu_0 \cdot \frac{\partial^2 \vec{E}}{\partial t^2} - \mu_0 \cdot \frac{\partial \vec{J}}{\partial t} = \nabla\left(\frac{\varrho}{\varepsilon_0}\right) - \nabla^2 \vec{E}$$

In the case of a space, free of electrical charge ($\varrho = 0$), we have

$$\nabla^2 \vec{E} - \varepsilon_0 \cdot \mu_0 \cdot \frac{\partial^2 \vec{E}}{\partial t^2} = \mu_0 \cdot \frac{\partial \vec{J}}{\partial t} \tag{4.78}$$

For time-harmonic dependence, this equation changes into (see also 4.50)

$$\nabla^2 \underline{\vec{E}} + \omega^2 \cdot \varepsilon_0 \cdot \mu_0 \cdot \underline{\vec{E}} = j \cdot \omega \cdot \mu_0 \cdot \underline{\vec{J}} \tag{4.79}$$

Equation (4.79) corresponds to the inhomogeneous wave Eq. (4.73). From both equations, it can be seen that

$$\frac{1}{\sqrt{\varepsilon_0 \cdot \mu_0}}$$

is the velocity of the electromagnetic wave. Since electromagnetic waves propagate in free space at the speed of light, the speed of light is

$$c_0 = \frac{1}{\sqrt{\varepsilon_0 \cdot \mu_0}}$$

(4.80)

The speed of light in a vacuum is $2,99792458 \cdot 10^{-8} \text{m/s}$. Thus for the absolute permittivity, ε_0 (see Eqs. (2.8 and 3.47) is

$$\varepsilon_0 = 8,8541878 \cdot 10^{-12} \frac{A \cdot s}{V \cdot m}$$

(4.81)

$$\varepsilon_0 = \frac{1}{c_0^2 \cdot \mu_0} = \frac{1}{(2,99792458 \cdot 10^{-8})^2 \cdot 4 \cdot \pi \cdot 10^{-7}} \cdot \frac{s^2 \cdot A \cdot m}{m^2 \cdot V \cdot s}$$

With (4.80), the wave Eq. (4.79) for the electric field vector is

$$\nabla^2 \vec{E} + \frac{\omega^2}{c_0^2} \cdot \vec{E} = j \cdot \omega \cdot \mu_0 \cdot \vec{J}$$

(4.82)

The origin for deriving the wave equation for the magnetic field vector is also the first Maxwell´s equation in the form of (4.75)

$$\nabla \times \vec{B} = \frac{1}{c_0^2} \cdot \frac{\partial \vec{E}}{\partial t} + \mu_0 \cdot \vec{J}$$

Applying another curl, we have

$$\nabla \times \left(\nabla \times \vec{B} \right) = \frac{1}{c_0^2} \cdot \frac{\partial}{\partial t} \left(\nabla \times \vec{E} \right) + \mu_0 \cdot \nabla \times \vec{J}$$

(4.83)

With a relationship of the vector analysis rules (see 3.88)

$$\nabla \times \left(\nabla \times \vec{B} \right) = \nabla \left(\nabla \cdot \vec{B} \right) - \nabla^2 \vec{B}$$

(4.84)

and Eq. (4.43)

$$\nabla \cdot \vec{B} = 0$$

we obtain with (4.83)

$$\frac{1}{c_0^2} \cdot \frac{\partial}{\partial t} \left(\nabla \times \vec{E} \right) + \mu_0 \cdot \nabla \times \vec{J} = -\nabla^2 \vec{B}$$

With the second Maxwell's Eq. (4.21)

$$\nabla \times \vec{E} = -\frac{\partial \vec{B}}{\partial t}$$

we get

$$-\frac{1}{c_0^2} \cdot \frac{\partial^2 \vec{B}}{\partial t^2} + \mu_0 \cdot \nabla \times \vec{J} = -\nabla^2 \vec{B}$$

or

$$\nabla^2 \vec{B} - \frac{1}{c_0^2} \cdot \frac{\partial^2 \vec{B}}{\partial t^2} = -\mu_0 \cdot \nabla \times \vec{J} \qquad (4.85)$$

Equation (4.85) corresponds to (4.73), the general inhomogeneous wave equation. The term $\left(\mu_0 \cdot \nabla \times \vec{J}\right)$ in (4.85) describes the excitation of the wave by the current density $\vec{J}$. For time-harmonic dependence, Eq. (4.85) can be written (see 4.50) as

$$\nabla^2 \underline{\vec{B}} + \frac{\omega^2}{c_0^2} \cdot \underline{\vec{B}} = -\mu_0 \cdot \nabla \times \underline{\vec{J}} \qquad (4.86)$$

4.10 Inhomogeneous Wave Equation for the Magnetic Vector Potential

It is useful to solve Maxwell's equations with the help of the magnetic vector potential. The integration of the partial differential equation of the vector potential, to be shown in this chapter, is easier than the direct integration of Maxwell's field equations. The magnetic and electric field vectors of interest can be obtained from the vector potential by simple differentiation.

According to (3.96), the relationship between the magnetic field $\underline{\vec{H}}$ and the vector potential $\underline{\vec{A}}$ for time-harmonic dependence is (seeing also Sect. 4.7)

$$\underline{\vec{B}} = \nabla \times \underline{\vec{A}} \qquad (4.87)$$

With (4.52)

$$\nabla \times \underline{\vec{E}} = -j \cdot \omega \cdot \underline{\vec{B}}$$

thus

$$\nabla \times \underline{\vec{E}} = -j \cdot \omega \cdot \left(\nabla \times \underline{\vec{A}}\right)$$

or

$$\nabla \times \left(\underline{\vec{E}} + j \cdot \omega \cdot \underline{\vec{A}}\right) = 0 \qquad (4.88)$$

With (3.86)

$$\nabla \times (\nabla \varphi) = 0$$

For time-harmonic dependence

$$\nabla \times \left(\nabla \underline{\varphi}\right) = 0$$

Together with (4.88), a scalar potential φ will be defined:

$$\underline{\vec{E}} + j \cdot \omega \cdot \underline{\vec{A}} = -\nabla \underline{\varphi} \qquad (4.89)$$

or

$$\underline{\vec{E}} = -\nabla \underline{\varphi} - j \cdot \omega \cdot \underline{\vec{A}} \qquad (4.90)$$

In contrast to the static case (see (1.5)), the so-defined scalar potential φ or $\underline{\varphi}$ depends both on the electric vector field $\underline{\vec{E}}$ and on the magnetic vector potential $\underline{\vec{A}}$

$$-\nabla \underline{\varphi} = \underline{\vec{E}} + j \cdot \omega \cdot \underline{\vec{A}}$$

If $\underline{\varphi}$ and $\underline{\vec{A}}$ are known, then $\underline{\vec{E}}$ and $\underline{\vec{B}}$ can be determined from Eqs. (4.87 and 4.90). Now the question is how the scalar potential $\underline{\varphi}$ and the vector potential $\underline{\vec{A}}$ can be calculated. For this purpose, (4.87) and (4.90) are combined with the still unused Maxwell's Eq. (4.51).

With (4.51)

$$\nabla \times \underline{\vec{H}} = \underline{\vec{J}} + j \cdot \omega \cdot \underline{\vec{D}}$$

we obtain

$$\nabla \times \underline{\vec{H}} = \frac{1}{\mu_0} \cdot \nabla \times \underline{\vec{B}} = \underline{\vec{J}} + j \cdot \omega \cdot \underline{\vec{D}} = \underline{\vec{J}} + j \cdot \omega \cdot \varepsilon_0 \cdot \underline{\vec{E}}$$

With the speed of light

$$c_0 = \frac{1}{\sqrt{\varepsilon_0 \cdot \mu_0}}$$

we have

$$\nabla \times \underline{\vec{B}} = \mu_0 \cdot \underline{\vec{J}} + j \cdot \omega \cdot \frac{1}{c_0^2} \cdot \underline{\vec{E}} \qquad (4.91)$$

By applying a relationship of the vector analysis (see 3.88)

$$\nabla^2 \underline{\vec{A}} = \nabla \left(\nabla \cdot \underline{\vec{A}}\right) - \nabla \times \left(\nabla \times \underline{\vec{A}}\right) \qquad (4.92)$$

we have with (4.87 and 4.91)

$$\nabla^2 \underline{\vec{A}} = \nabla \left(\nabla \cdot \underline{\vec{A}} \right) - \mu_0 \cdot \underline{\vec{J}} - j \cdot \omega \cdot \frac{1}{c_0^2} \underline{\vec{E}} \tag{4.93}$$

$$\nabla^2 \underline{\vec{A}} = \nabla \left(\nabla \cdot \underline{\vec{A}} \right) - \nabla \times \vec{B}$$

Inserting $\vec{E}$ from (4.90), we get

$$\nabla^2 \underline{\vec{A}} = \nabla \left(\nabla \cdot \underline{\vec{A}} \right) - \mu_0 \cdot \underline{\vec{J}} - j\omega \cdot \frac{1}{c_0^2} \left(-\nabla \underline{\varphi} - j\omega \cdot \underline{\vec{A}} \right)$$

or

$$\nabla^2 \underline{\vec{A}} = \nabla \left(\nabla \cdot \underline{\vec{A}} + j\omega \cdot \frac{1}{c_0^2} \cdot \underline{\varphi} \right) - \mu_0 \cdot \underline{\vec{J}} - \frac{\omega^2}{c_0^2} \cdot \underline{\vec{A}}$$

and

$$\nabla^2 \underline{\vec{A}} + \frac{\omega^2}{c_0^2} \cdot \underline{\vec{A}} - \nabla \left(\nabla \cdot \underline{\vec{A}} + \frac{j\omega}{c_0^2} \cdot \underline{\varphi} \right) = -\mu_0 \cdot \underline{\vec{J}} \tag{4.94}$$

respectively.

This relationship can be simplified with Helmholtz' theorem[10]. The theorem says[11]:

A vector field is completely determined by its sources and its curls, except for an additive constant.

So far, by (4.87) only the curls of the vector field were determined by $\left(\underline{\vec{B}} = \nabla \times \underline{\vec{A}} \right)$. The sources are still freely available. The sources are conveniently defined so that Eq. (4.94) is as simple as possible, i.e., the parenthesis expression in Eq. (4.94) is set to zero

$$\nabla \cdot \underline{\vec{A}} + \frac{j\omega}{c_0^2} \cdot \underline{\varphi} = 0 \tag{4.95}$$

This agreement is called Lorentz calibration. With Eqs. (4.95 and 4.94) changes into

$$\nabla^2 \underline{\vec{A}} + \frac{\omega^2}{c_0^2} \cdot \underline{\vec{A}} = -\mu_0 \cdot \underline{\vec{J}} \tag{4.96}$$

Equation (4.96) has the form of (4.73) and is the inhomogeneous wave equation for the vector potential $\underline{\vec{A}}$ in the case of time-harmonic dependence. With (4.90)

$$\underline{\vec{E}} = -\nabla \underline{\varphi} - j \cdot \omega \cdot \underline{\vec{A}}$$

---

[10]Helmholtz, Hermann Ludwig Ferdinand, from 1883 von Helmholtz, German physiologist and physicist, * 1821, † 1894.

[11]See [2] and [5] and the remark in [3] on p. 322.

or

$$\nabla \underline{\varphi} = -\underline{\vec{E}} - j \cdot \omega \cdot \underline{\vec{A}}$$

and with (4.95), we obtain

$$\nabla \left(\nabla \cdot \underline{\vec{A}} \right) + \frac{j\omega}{c_0^2} \cdot \nabla \underline{\varphi} = 0$$

and also

$$\nabla \left(\nabla \cdot \underline{\vec{A}} \right) + \frac{j\omega}{c_0^2} \cdot \left(-\underline{\vec{E}} - j \cdot \omega \cdot \underline{\vec{A}} \right) = 0$$

$$\frac{j\omega}{c_0^2} \cdot \underline{\vec{E}} = \nabla \left(\nabla \cdot \underline{\vec{A}} \right) + \frac{\omega^2}{c_0^2} \cdot \underline{\vec{A}}$$

$$\underline{\vec{E}} = \frac{c_0^2}{j\omega} \cdot \left(\nabla \left(\nabla \cdot \underline{\vec{A}} \right) + \frac{\omega^2}{c_0^2} \cdot \underline{\vec{A}} \right) \tag{4.97}$$

Thus for a known magnetic vector potential $\vec{A}$, the electric field can be calculated and the magnetic field with (4.87). Alternatively, the electric field $\underline{\vec{E}}$ can be calculated with (4.51)

$$\underline{\vec{E}} = \frac{1}{j \cdot \omega \cdot \varepsilon_0} \left(\nabla \times \underline{\vec{H}} - \underline{\vec{J}} \right) \tag{4.98}$$

4.11   Solution of the Wave Equation for the Vector Potential by the Retarded Potential

The integration of the inhomogeneous wave Eq. (4.96) is a wave emanating from a transmitter. This wave is caused by a time- and space-dependent current density $\vec{J}$. The integration of the partial differential equation for the vector potential is simpler than the direct integration of Maxwell's field equations. The calculation of the vector potential for the static case, according to Eq. (3.103), is a special case of the solution of the wave Eq. (4.96). Analogous to the static case, the vector potential $\underline{\vec{A}}(\vec{r}, t)$, at the time t at the location P or $\vec{r}$, is a superposition of the contributions of the time-dependent current densities $\underline{\vec{J}}(\vec{r})$ (see Fig. 4.21). It should be noted that the contribution generated by the volume element dV at time t in point P has the distance $|\vec{r} - \vec{r}_V|$ from point P. It was thus transmitted at the time

$$t^* = t - \frac{|\vec{r} - \vec{r}_V|}{c_0} \tag{4.99}$$

Fig. 4.21 Time- and location-dependent vector field of the current density $\vec{\underline{J}}$

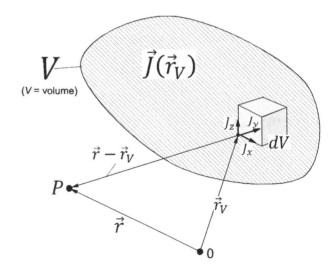

i.e., $|\vec{r} - \vec{r}_V|/c_0$ earlier than t. Free space is assumed to be the propagation medium. The propagation velocity is the velocity of light c_0 and therefore $\mu = \mu_0$ and $\varepsilon = \varepsilon_0$.

To solve the partial differential Eq. (4.96), the Eq. (3.103), which is valid for the static case, is to be modified

$$\vec{A}(\vec{r},t) = \frac{\mu_0}{4 \cdot \pi} \cdot \iiint\limits_{V} \frac{\vec{J}(\vec{r}_V, t^*)}{|\vec{r} - \vec{r}_V|} \cdot \mathrm{d}V' \qquad (4.100)$$

The vector potential in Eq. (4.100) is called retarded potential, because it is generated by contributions that are emitted at the so-called retarded time t^*. For the case of the time-harmonic dependence of the current density in volume element $\mathrm{d}V$ according to (4.48)

$$\vec{J}(\vec{r}_V, t^*) = Re\left\{ \vec{J}(\vec{r}) \cdot e^{j \cdot \varphi_0} \cdot e^{j \cdot \omega \cdot \left(t - \frac{|\vec{r} - \vec{r}_V|}{c_0}\right)} \right\}$$

with

$$\underline{\vec{J}}(\vec{r}) = \vec{J}(\vec{r}) \cdot e^{j \cdot \varphi_0} \cdot e^{j \cdot \omega \cdot t}$$

we have

$$\vec{J}(\vec{r}_V, t^*) = Re\left\{ \vec{J}(\vec{r}) \cdot e^{j \cdot \varphi_0} \cdot e^{j \cdot \omega \cdot t} \cdot e^{j \cdot \omega \cdot \left(-\frac{|\vec{r} - \vec{r}_V|}{c_0}\right)} \right\} = Re\left\{ \underline{\vec{J}}(\vec{r}) \cdot e^{-j \cdot \frac{\omega}{c_0} \cdot |\vec{r} - \vec{r}_V|} \right\}$$

As described in Sect. 4.7.1 for time-harmonic dependence, the calculation is performed with the complex vector $\underline{J}(\vec{r})$ including the factor $e^{j \cdot \omega \cdot t}$. For the case of time-harmonic dependence, the current density $\vec{J}(\vec{r}', t^*)$ in (4.100) is to be replaced consequently by

$$\underline{J}(\vec{r}) \cdot e^{-j \cdot \frac{\omega}{c_0} \cdot |\vec{r} - \vec{r}_V|}$$

With (4.100), we then have

$$\vec{\underline{A}}(\vec{r}, t) = \frac{\mu_0}{4 \cdot \pi} \cdot \iiint\limits_V \vec{\underline{J}}(\vec{r}) \cdot \frac{e^{-j \cdot \frac{\omega}{c_0} \cdot |\vec{r} - \vec{r}_V|}}{|\vec{r} - \vec{r}_V|} \cdot dV \qquad (4.101)$$

In Chap. 5, this equation is the starting point for calculating the electromagnetic waves emanating from the Hertzian dipole.

4.12 Energy Transport in the Electromagnetic Field

The equations for calculating the energy density w_{el} stored in the static electric fields and the energy density w_{magn} stored in the static magnetic field have been derived in Chaps. 1 and 2 (see 2.59 and 4.15)

$$w_{el} = \frac{1}{2} \cdot \varepsilon \cdot E^2 \qquad (4.102)$$

and

$$w_{magn} = \frac{1}{2} \cdot \mu \cdot H^2 \qquad (4.103)$$

Figure 4.22 shows a closed surface A of volume V. To represent the energy flux through the surface A, a vector $\vec{S}$ is introduced, which represents the electromagnetic energy density, i.e., the energy per surface element dA moving into the direction of the vector $\vec{S}$[12]. This vector is called Poynting vector.

The Poynting vector points into the spatial direction of the energy flux. Its magnitude corresponds to the power density of the wave, i.e., the energy that passes through a surface element per unit time. This surface element vector points in the same direction as the Poynting vector. The Poynting vector is named after the English physicist who introduced the concept of energy flux to electrodynamics[13]. The Poynting vector has the dimension

---

[12]The concept of energy flow is identical with the physical concept of power. The term energy flux density is therefore equivalent to power density.

[13]John Henry Poynting, English physicist, *1852, †1914.

Fig. 4.22 Poynting vector

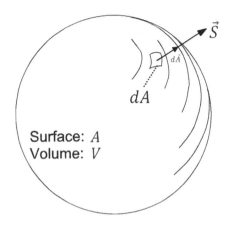

Surface: A
Volume: V

$$\frac{\text{Energy}}{\text{Surface} \cdot \text{Time}} = \frac{\text{Power}}{\text{Surface}}$$

Since the vector $\vec{S}$ indicates the energy flux density, the power P that flows through the closed surface A out of the volume V is

$$P = \oiint_A \vec{S} \cdot d\vec{A} = -\iiint_V \frac{dW}{dt} \cdot dV \tag{4.104}$$

In this equation is dW/dt the energy flowing out of the volume element dV per unit time. With the integral theorem of Gauss (see 2.35)

$$P = \oiint_A \vec{S} \cdot d\vec{A} = -\iiint_V \left(\nabla \cdot \vec{S}\right) \cdot dV = -\iiint_V \frac{dW}{dt} \cdot dV \tag{4.105}$$

The energy flowing out of the volume element dV per unit time, or the energy flux in the electromagnetic field (compare 4.28), is

$$-\frac{dW}{dt} = \nabla \cdot \vec{S} \tag{4.106}$$

The negative sign in this equation indicates that an energy flux in the direction of the Poynting vector reduces the energy density in the volume. It must be possible to express the energy flux, represented by the Poynting vector, by the field quantities of the electromagnetic field. For this verification, we start with Maxwell's Eqs. (4.38 and 4.21).

First Maxwell's equation

$$\nabla \times \vec{H} = \frac{\partial \vec{D}}{\partial t} + \vec{J}$$

Second Maxwell's equation

$$\nabla \times \vec{E} = -\frac{\partial \vec{B}}{\partial t}$$

According to (2.59) and (4.15) the energy density is proportional to the square of the magnitude of the electric resp. magnetic field. For this reason, as a first step, the first Maxwell's equation is multiplied by $\vec{E}$ and the second Maxwell's equation by $\vec{H}$. Then, these equations will be subtracted

$$\vec{H} \cdot \left(\nabla \times \vec{E} \right) - \vec{E} \cdot \left(\nabla \times \vec{H} \right) = -\mu \cdot \vec{H} \cdot \frac{\partial \vec{H}}{\partial t} - \varepsilon \cdot \vec{E} \cdot \frac{\partial \vec{E}}{\partial t} - \vec{E} \cdot \vec{J}$$

With (1.22)

$$\sigma \cdot \vec{E} = \vec{J} \tag{4.107}$$

where $\sigma = $ *specific conductivity*

$$\vec{H} \cdot \left(\nabla \times \vec{E} \right) - \vec{E} \cdot \left(\nabla \times \vec{H} \right) = -\mu \cdot \vec{H} \cdot \frac{\partial \vec{H}}{\partial t} - \varepsilon \cdot \vec{E} \cdot \frac{\partial \vec{E}}{\partial t} - \sigma \cdot \vec{E} \cdot \vec{E} \tag{4.108}$$

With rules of the vector analysis (see 3.89)

$$\nabla \cdot \left(\vec{E} \times \vec{H} \right) = \vec{H} \cdot \left(\nabla \times \vec{E} \right) - \vec{E} \cdot \left(\nabla \times \vec{H} \right)$$

From (4.108)

$$\nabla \cdot \left(\vec{E} \times \vec{H} \right) = -\mu \cdot \vec{H} \cdot \frac{\partial \vec{H}}{\partial t} - \varepsilon \cdot \vec{E} \cdot \frac{\partial \vec{E}}{\partial t} - \sigma \cdot \vec{E} \cdot \vec{E} \tag{4.109}$$

In (4.109), $\sigma \cdot \vec{E} \cdot \vec{E}$ is the part of the energy converted into heat by the current flow, i.e., it determines the loss of electromagnetic energy. For a non-conductive medium $\sigma = 0$

$$\nabla \cdot \left(\vec{E} \times \vec{H} \right) = -\mu \cdot \vec{H} \cdot \frac{\partial \vec{H}}{\partial t} - \varepsilon \cdot \vec{E} \cdot \frac{\partial \vec{E}}{\partial t} \tag{4.110}$$

or

$$\nabla \cdot \left(\vec{E} \times \vec{H} \right) = -\frac{\partial}{\partial t} \cdot \left(\frac{1}{2} \cdot \mu \cdot H^2 + \frac{1}{2} \cdot \varepsilon \cdot E^2 \right) \tag{4.111}$$

According to (4.102 and 4.103), the expression in the parenthesis of Eq. (4.111) is the energy density present in the magnetic and electric field. The whole expression in this equation is thus the energy flux in the electromagnetic field. According to (4.111 and 4.106), the Poynting vector is

$$\vec{S} = \vec{E} \times \vec{H} \tag{4.112}$$

In the case of the time-harmonic dependence of $\vec{E}$ and $\vec{H}$, according to (4.48), the complex field vectors $\underline{\vec{E}}$ and $\underline{\vec{H}}$ will be used. The relationship between the Poynting vector and the energy flux density, for this case, is derived from the analogy of the calculation of the power from the phasor representation of the voltage and the current for the case of time-harmonic dependence.

According to (4.47), a cosinusoidal voltage and current are

$$V(t) = V \cdot \cos(\omega \cdot t + \varphi_v) \tag{4.113}$$

and

$$I(t) = I \cdot \cos(\omega \cdot t + \varphi_i) \tag{4.114}$$

In these equations, V and I are the voltage and current amplitudes, and φ_v and φ_i are the phase angles. The phasor representations of $V(t)$ and $I(t)$ are

$$\underline{V}(t) = V \cdot e^{j \cdot (\omega \cdot t + \varphi_v)} \tag{4.115}$$

and

$$\underline{I}(t) = I \cdot e^{j \cdot (\omega \cdot t + \varphi_i)} \tag{4.116}$$

The real part of $\underline{V}(t)$ and $\underline{I}(t)$ in (4.113 and 4.114) gives voltage and current

$$V(t) = Re\{V \cdot e^{j \cdot (\omega \cdot t + \varphi_v)}\} = Re\{V \cdot \cos(\omega \cdot t + \varphi_v) + j \cdot V \cdot \sin(\omega \cdot t + \varphi_v)\} \tag{4.117}$$
$$I(t) = Re\{I \cdot e^{j \cdot (\omega \cdot t + \varphi_i)}\} = Re\{I \cdot \cos(\omega \cdot t + \varphi_i) + j \cdot I \cdot \sin(\omega \cdot t + \varphi_i)\} \tag{4.118}$$

The power P is the product of voltage and current

$$P(t) = V(t) \cdot I(t) = V \cdot \cos(\omega \cdot t + \varphi_v) \cdot I \cdot \cos(\omega \cdot t + \varphi_i)$$

With the addition theorem

$$\cos\alpha \cdot \cos\beta = \frac{1}{2} \cdot [\cos(\alpha - \beta) + \cos(\alpha + \beta)]$$

we have

$$P(t) = \frac{V \cdot I}{2} \cdot [\cos(\varphi_v - \varphi_i) + \cos(2 \cdot \omega \cdot t + \varphi_v + \varphi_i)] \tag{4.119}$$

In (4.119), the first summand is the time-independent power

$$P = \frac{V \cdot I}{2} \cdot \cos(\varphi_v - \varphi_i) \tag{4.120}$$

With

$$\cos(-\alpha) = \cos(\alpha)$$

we obtain

$$P = \frac{V \cdot I}{2} \cdot \cos (\varphi_i - \varphi_v) \tag{4.121}$$

The second summand in (4.119) does not contribute to the power. This part, integrated over the period T, is equal to zero

$$\frac{V \cdot I}{2} \cdot \int_0^T \cos \left(2 \cdot \frac{2 \cdot \pi}{T} \cdot t + \varphi_v + \varphi_i \right) dt = \frac{V \cdot I}{2} \cdot \cos (\varphi_v + \varphi_i) \cdot \int_0^T \cos \left(\frac{4 \cdot \pi}{T} \cdot t \right) dt$$

$$- \frac{V \cdot I}{2} \cdot \sin (\varphi_v + \varphi_i) \cdot \int_0^T \sin \left(\frac{4 \cdot \pi}{T} \cdot t \right) dt$$

$$\int_0^T \cos \left(\frac{4 \cdot \pi}{T} \cdot t \right) dt = \frac{T}{4 \cdot \pi} \cdot \left[\sin \left(\frac{4 \cdot \pi}{T} \cdot t \right) \right]_0^T = \frac{T}{4 \cdot \pi} \cdot [\sin (4 \cdot \pi) - \sin (0)] = 0$$

$$\int_0^T \sin \left(\frac{4 \cdot \pi}{T} \cdot t \right) dt = -\frac{T}{4 \cdot \pi} \cdot \left[\cos \left(\frac{4 \cdot \pi}{T} \cdot t \right) \right]_0^T = -\frac{T}{4 \cdot \pi} \cdot [\cos (4 \cdot \pi) - \cos (0)]$$

$$= -\frac{T}{4 \cdot \pi} (1 - 1) = 0$$

The power in (4.120) can also be expressed in form of a phasor. From (4.115 and 4.116)

$$\underline{V}(t) \cdot \underline{I}(t) = V \cdot I \cdot e^{j \cdot (\omega \cdot t + \varphi_v)} \cdot e^{j \cdot (\omega \cdot t + \varphi_i)}$$

With the current

$$\underline{I}(t)^* = I \cdot e^{-j \cdot (\omega \cdot t + \varphi_i)}$$

complex conjugate to

$$\underline{I}(t) = I \cdot e^{j \cdot (\omega \cdot t + \varphi_i)}$$

we have[14]

$$\underline{V}(t) \cdot \underline{I}(t)^* = V \cdot I \cdot \left[\cos (\varphi_v - \varphi_i) + j \cdot \sin (\varphi_v - \varphi_i) \right] \tag{4.122}$$

$$\underline{V}(t) \cdot \underline{I}(t)^* = V \cdot I \cdot e^{j \cdot (\omega \cdot t + \varphi_v)} \cdot e^{-j \cdot (\omega \cdot t + \varphi_i)}$$

$$\underline{V}(t) \cdot \underline{I}(t)^* = V \cdot I \cdot e^{j \cdot (\varphi_v - \varphi_i)}$$

The real part divided by 2 is

$$\frac{1}{2} \cdot Re\{\underline{V} \cdot \underline{I}^*\} = \frac{V \cdot I}{2} \cdot \cos (\varphi_v - \varphi_i)$$

---

[14]The * indicates the complex conjugated value.

And with (4.120 and 4.121)

$$P = \frac{1}{2} \cdot Re\{\underline{V} \cdot \underline{I}^*\} = \frac{1}{2} \cdot Re\{\underline{V}^* \cdot \underline{I}\}$$

(4.123)

Equation (4.123) can be applied analogously to the vector product of the electric field vector $\vec{E}$ and on the magnetic field vector $\vec{H}$. Therefore, the active power density transported by the electromagnetic field, into the direction of the Poynting vector, for time-harmonic dependence, is

$$\underline{\vec{S}} = \frac{1}{2} \cdot Re\{\underline{\vec{E}} \times \underline{\vec{H}}^*\} = \frac{1}{2} \cdot Re\{\underline{\vec{E}}^* \times \underline{\vec{H}}\}$$

(4.124)

Wave Propagation

5

In the previous section, the wave equations, which are implicitly included in Maxwell's equations, were derived and their solution given by the help of the magnetic vector potential. Electromagnetic waves can propagate along lines either in the form of a single-wire, a two-wire line, a coaxial line, a stripline, or within waveguides. But even without such transmission lines, electromagnetic waves can propagate in free space. The equations for the propagation of electromagnetic waves on transmission lines are based on the boundary conditions at the surfaces carrying the wave. The propagation in free space, on the other hand, is determined by the type of excitation by the transmitter and its transmitting antenna. The propagation of electromagnetic waves along transmission lines is not the subject of this volume. The solutions of Maxwell's equations for these applications are discussed in detail, e.g., [2, 4], and [6].

For simplicity, the wave propagation in free space was chosen with wave excitation by a Hertzian dipole. Maxwell's equations can easily be solved using this example. The electromagnetic field can be calculated and represented with little effort and the essential characteristics of an antenna can be derived.

5.1 Hertzian Dipole

The Hertzian dipole, also called the electrical elementary dipole, is a transmitter antenna consisting of a wire-shaped conductor, short in relation to the wavelength, and with a current density constant along the length of the wire. The current is assumed to be cosinusoidal with time. A Hertzian dipole is visualized in Fig. 5.1 as the transformation from a very small plate capacitor. The alternating current source is connected to the terminals of the Hertzian dipole by a twisted two-wire transmission line or a coaxial cable, so that

© Springer Fachmedien Wiesbaden GmbH, part of Springer Nature 2020
J. Donnevert, *Maxwell's Equations*, https://doi.org/10.1007/978-3-658-29376-5_5

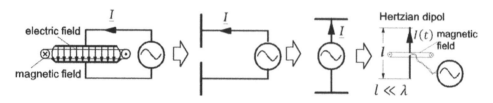

Fig. 5.1 Realization of the Hertzian dipole

an electromagnetic wave is prevented from being radiated from the feeding line or the cable[1].

The current as a function of time within the Hertzian dipole is assumed as

$$I(t) = I \cdot \cos(\omega \cdot t + \varphi_0) = \text{Re}\left(I \cdot e^{j \cdot \varphi_0} \cdot e^{j \cdot \omega \cdot t}\right) = \text{Re}\left(\underline{I} \cdot e^{j \cdot \omega \cdot t}\right) \qquad (5.1)$$

If dA is the cross section of the conductor of the Hertzian dipole, the current density in complex notation (see (4.48)) is

$$\underline{J}(t) = \frac{\underline{I}(t)}{dA} \qquad (5.2)$$

The underscores in (5.2) indicate that current and current density have a time-harmonic dependence. A constant phase angle can be set to zero without limiting the general validity.

In Fig. 5.2, the Hertzian dipole is positioned at the origin of a spherical coordinate system. The current $\underline{I}(t)$ and thus also the current density $\underline{J}(t)$ are oriented into the z-direction. Consequently, the vector potential $\vec{A}(\vec{r}, t)$ has only a z-component.

A volume element of the Hertzian dipole is

$$dV = dA \cdot dz$$

The vector potential $\vec{A}_z$ of the Hertzian dipole of length l, located in the origin of the coordinate system ($\vec{r}_V = 0$), therefore is (see (4.101))

$$\underline{\vec{A}}_z(\vec{r}, t) = \frac{\mu_0}{4 \cdot \pi} \cdot \int\limits_{z=-\frac{l}{2}}^{z=+\frac{l}{2}} \frac{\underline{I}(t) \cdot \vec{e}_z}{dA} \cdot \frac{e^{-j \cdot \frac{\omega}{c_0} \cdot |\vec{r}|}}{|\vec{r}|} \cdot dA \cdot dz \qquad (5.3)$$

With $|\vec{r}| = r$

$$\underline{\vec{A}}_z(\vec{r}, t) = \frac{\mu_0}{4 \cdot \pi} \cdot \left[\underline{I}(t) \cdot \vec{e}_z \cdot \frac{e^{-j \cdot \frac{\omega}{c_0} \cdot r}}{r} \cdot z \right]_{z=-\frac{l}{2}}^{z=+\frac{l}{2}} \qquad (5.4)$$

---

[1][6, p. 507 ff.].

Fig. 5.2 Hertzian dipole at the origin of the spherical coordinate system

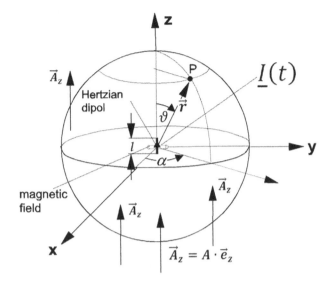

$$\underline{\vec{A}}_z(\vec{r}, t) = \left(\frac{\mu_0}{4 \cdot \pi} \cdot \underline{I}(t) \cdot l \cdot \frac{e^{-j \cdot \frac{\omega}{c_0} \cdot r}}{r} \right) \cdot \vec{e}_z \qquad (5.5)$$

Thus with (3.96)

$$\underline{\vec{B}} = \mu_0 \cdot \underline{\vec{H}} = \nabla \times \underline{\vec{A}}$$

we have

$$\underline{\vec{H}}(\vec{r}, t) = \frac{1}{\mu_0} \nabla \times \left[\underline{\vec{A}}_z(\vec{r}, t) \right] = \frac{1}{\mu_0} \nabla \times \left[\underline{A}_z(\vec{r}, t) \cdot \vec{e}_z \right] \qquad (5.6)$$

Since $\underline{\vec{A}}_z(\vec{r}, t)$ depends on the radius $\vec{r}$, the unit vector $\vec{e}_z$ is converted into the unit vectors $\vec{e}_r$ and $\vec{e}_\vartheta$ of the spherical coordinate system.

$$\vec{e}_z = \vec{e}_r \cdot \cos \vartheta - \vec{e}_\vartheta \cdot \sin \vartheta \qquad (5.7)$$

This relationship is shown in Fig. 5.3. Thus

$$\begin{aligned} \underline{\vec{H}}(\vec{r}, t) &= \frac{1}{\mu_0} \cdot \nabla \times \left[\underline{\vec{A}}_z(\vec{r}, t) \right] \\ &= \frac{1}{\mu_0} \cdot \nabla \times \left[\vec{e}_r \cdot \underline{A}_z(\vec{r}, t) \cdot \cos \vartheta - \vec{e}_\vartheta \cdot \underline{A}_z(\vec{r}, t) \cdot \sin \vartheta \right] \end{aligned} \qquad (5.8)$$

With (5.5)

$$\underline{\vec{H}}(\vec{r}, t) = \frac{\underline{I}(t) \cdot l}{4 \cdot \pi} \nabla \times \left[\vec{e}_r \cdot \left(\frac{e^{-j \cdot \frac{\omega}{c_0} \cdot r}}{r} \right) \cdot \cos \vartheta - \vec{e}_\vartheta \cdot \left(\frac{e^{-j \cdot \frac{\omega}{c_0} \cdot r}}{r} \right) \cdot \sin \vartheta \right] \qquad (5.9)$$

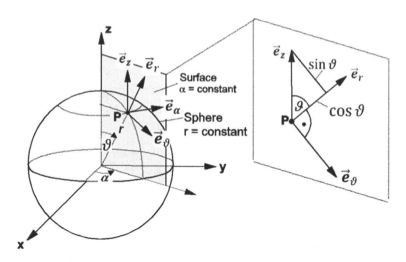

Fig. 5.3 Conversion of the unit vector $\vec{e}_z$ into the unit vectors $\vec{e}_r$ and $\vec{e}_\vartheta$

The curl of the vector potential $\vec{A}$, i.e., $(\nabla \times \vec{A})$ in spherical coordinates (cf. (3.85))

$$\left(\nabla \times \vec{A}\right) = \vec{e}_r \cdot \frac{1}{r \cdot \sin \vartheta} \cdot \left[\frac{\partial}{\partial \vartheta}(A_\alpha \cdot \sin \vartheta) - \frac{\partial A_\vartheta}{\partial \alpha}\right]$$

$$+ \vec{e}_\vartheta \cdot \frac{1}{r} \cdot \left[\frac{1}{\sin \vartheta} \frac{\partial A_r}{\partial \alpha} - \frac{\partial}{\partial r}(r \cdot A_\alpha)\right] \tag{5.10}$$

$$+ \vec{e}_\alpha \cdot \frac{1}{r} \cdot \left[\frac{\partial}{\partial r}(r \cdot A_\vartheta) - \frac{\partial A_r}{\partial \vartheta}\right]$$

The vector potential $\vec{A}$ in (5.8) has only an r- and an ϑ-component. Both components depend only on r and ϑ and not on α. As a result, the magnetic field, according to (5.10), has only an α-component, i.e.,

$$\underline{H}_r(\vec{r}, t) = 0 \text{ and } \underline{H}_\vartheta(\vec{r}, t) = 0$$

Equation (5.9) thus changes into

$$\vec{\underline{H}}(\vec{r}, t) = \underline{H}_\alpha(\vec{r}, t) \cdot \vec{e}_\alpha$$

$$= \frac{I(t) \cdot l}{4 \cdot \pi} \cdot \frac{1}{r} \left[-\frac{\partial}{\partial \vartheta}\left(\frac{e^{-j \cdot \frac{\omega}{c_0} \cdot r}}{r} \cdot \cos \vartheta\right) - \frac{\partial}{\partial r}\left(e^{-j \cdot \frac{\omega}{c_0} \cdot r} \cdot \sin \vartheta\right)\right] \cdot \vec{e}_\alpha$$

$$\tag{5.11}$$

$$\underline{H}_\alpha(\vec{r}, t) = \frac{I(t) \cdot l}{4 \cdot \pi} \cdot \frac{1}{r} \left[\frac{e^{-j \cdot \frac{\omega}{c_0} \cdot r}}{r} \cdot \sin \vartheta - \sin \vartheta \cdot \left(-j \cdot \frac{\omega}{c_0}\right) \cdot e^{-j \cdot \frac{\omega}{c_0} \cdot r}\right]$$

$$\underline{H}_\alpha(\vec{r}, t) = \frac{I(t) \cdot l}{4 \cdot \pi} \cdot \frac{1}{r} \cdot \left[\frac{e^{-j \cdot \frac{\omega}{c_0} \cdot r}}{r} \cdot \sin \vartheta + \sin \vartheta \cdot \left(j \cdot \frac{\omega}{c_0} \right) \cdot e^{-j \cdot \frac{\omega}{c_0} \cdot r} \right]$$

$$\underline{H}_\alpha(\vec{r}, t) = j \cdot I(t) \cdot l \cdot \frac{\omega}{c_0} \cdot \sin \vartheta \cdot \frac{e^{-j \cdot \frac{\omega}{c_0} \cdot r}}{4 \cdot \pi \cdot r} \cdot \left[1 + \frac{c_0}{j \cdot \omega \cdot r} \right] \tag{5.12}$$

The wavelength λ^2 is

$$\lambda = \frac{c_0}{f} = \frac{c_0 \cdot 2 \cdot \pi}{\omega} \tag{5.13}$$

with

$$\frac{\omega}{c_0} = \frac{2 \cdot \pi}{\lambda} \tag{5.14}$$

With (5.12) and the wavelength (5.14)

$$\vec{H}(\vec{r}, t) = j \cdot \frac{I(t) \cdot l}{2 \cdot \lambda} \cdot \frac{\sin \vartheta}{r} \cdot e^{-j \cdot \frac{2 \cdot \pi}{\lambda} \cdot r} \cdot \left[1 + \frac{1}{j \cdot \frac{2 \cdot \pi \cdot r}{\lambda}} \right] \vec{e}_\alpha$$

$$\underline{\vec{H}}(\vec{r}, t) = j \cdot \frac{I(t) \cdot l \cdot \pi}{\lambda^2} \cdot \sin \vartheta \cdot \frac{e^{-j \cdot \frac{2 \cdot \pi}{\lambda} \cdot r}}{\frac{2 \cdot \pi}{\lambda} \cdot r} \cdot \left[1 + \frac{1}{j \cdot \frac{2 \cdot \pi \cdot r}{\lambda}} \right] \vec{e}_\alpha \tag{5.15}$$

The expression

$$\frac{\pi \cdot I(t) \cdot l}{\lambda^2} = \underline{H}_0 \tag{5.16}$$

in (5.15) can be seen as the complex amplitude of the magnetic field. Thus from (5.15)

$$\underline{\vec{H}}(\vec{r}, t) = j \cdot \underline{H}_0 \cdot \sin \vartheta \cdot \frac{e^{-j \cdot \frac{2 \cdot \pi}{\lambda} \cdot r}}{\frac{2 \cdot \pi}{\lambda} \cdot r} \cdot \left[1 + \frac{1}{j \cdot \frac{2 \cdot \pi \cdot r}{\lambda}} \right] \vec{e}_\alpha \tag{5.17}$$

The ratio

$$\frac{2 \cdot \pi}{\lambda} = \frac{\omega}{c_0} = k_0 \tag{5.18}$$

is called wave number.

---

[2]The wavelength λ is the distance by which a phase state (e.g., a maximum or a zero crossing) of a wave progresses or propagates within a period of time T (see Fig. 4.19).

$T = 1/f$, $f =$ frequency, $f = \frac{\omega}{(2 \cdot \pi)}$, i.e., $\omega = 2 \cdot \pi \cdot f$, $\omega =$ angular frequency,

With (5.18), Eq. (5.17) can be further simplified

$$\underline{\vec{H}}(\vec{r},t) = j \cdot \underline{H}_0 \cdot \sin \vartheta \cdot \frac{e^{-j \cdot k_0 \cdot r}}{k_0 \cdot r} \cdot \left[1 + \frac{1}{j \cdot k_0 \cdot r}\right] \vec{e}_\alpha \qquad (5.19)$$

Equation (5.19) is the equation of a wave propagating into the positive r-direction. This can be seen if (5.19) is multiplied, according to (4.48), with the factor $e^{j \cdot \omega \cdot t}$ and then the real part is taken. This restores the time-harmonic dependence in the equation

$$\vec{H}(\vec{r},t) = \mathrm{Re}\left\{ j \cdot \underline{H}_0 \cdot \sin \vartheta \cdot \frac{e^{-j \cdot k_0 \cdot r}}{k_0 \cdot r} \cdot \left[1 + \frac{1}{j \cdot k_0 \cdot r}\right] \cdot e^{j \cdot \omega \cdot t} \right\} \vec{e}_\alpha$$

$$\vec{H}(\vec{r},t) = \mathrm{Re}\left\{ j \cdot \underline{H}_0 \cdot \sin \vartheta \cdot \frac{1}{k_0 \cdot r} \cdot \left[1 + \frac{1}{j \cdot k_0 \cdot r}\right] \cdot e^{j \cdot (\omega \cdot t - k_0 \cdot r)} \right\} \vec{e}_\alpha \qquad (5.20)$$

The exponent of the e-function (see (5.14) and (5.18))

$$(\omega \cdot t - k_0 \cdot r) = \omega \cdot \left(t - \frac{r}{c_0}\right)$$

is characteristic for wave propagation in the positive r-direction (see (4.55)).

After deriving the magnetic field vector with the help of the magnetic vector potential, the relationship for the vector of the electric field of the Hertzian dipole can be derived from Maxwell's first equation. According to (4.39), the first Maxwell's equation is

$$\nabla \times \vec{H} = \frac{\partial \vec{D}}{\partial t} + \vec{J} \qquad (5.21)$$

Since the current density $\vec{J}$ equals zero outside the Hertzian dipole

$$\nabla \times \vec{H} = \varepsilon_0 \frac{\partial \vec{E}}{\partial t}$$

For time-harmonic dependence

$$\nabla \times \vec{H} = \varepsilon_0 \cdot j \cdot \omega \cdot \vec{E} \qquad (5.22)$$

with (5.14), (5.18), and (4.80)

$$\frac{1}{\omega} = \frac{\lambda}{2 \cdot \pi \cdot c_0} = \frac{1}{c_0 \cdot k_0} = \frac{\sqrt{\varepsilon_0 \cdot \mu_0}}{k_0} \qquad (5.23)$$

and putting this relationship into (5.22)

$$\underline{\vec{E}} = \frac{1}{\varepsilon_0 \cdot j \cdot \omega} \cdot \left(\nabla \times \underline{\vec{H}}\right) = \frac{1}{j \cdot k_0} \cdot \frac{\sqrt{\varepsilon_0 \cdot \mu_0}}{\varepsilon_0} \cdot \left(\nabla \times \underline{\vec{H}}\right) = \frac{1}{j \cdot k_0} \cdot \sqrt{\frac{\mu_0}{\varepsilon_0}} \cdot \left(\nabla \times \underline{\vec{H}}\right)$$

$$(5.24)$$

In this equation

$$\sqrt{\frac{\mu_0}{\varepsilon_0}} = Z_0 \tag{5.25}$$

Z_0 is referred to as wave impedance Z_0 of the electromagnetic field[3]. From (5.24) and (5.25)

$$\vec{\underline{E}} = \frac{Z_0}{j \cdot k_0} \cdot \left(\nabla \times \vec{\underline{H}} \right) \tag{5.26}$$

According to (5.19), the magnetic field has only an α-component. Thus (5.26)

$$\vec{\underline{E}} = \frac{Z_0}{j \cdot k_0} \cdot \nabla \times \left(\underline{H}_\alpha \cdot \vec{e}_\alpha \right) \tag{5.27}$$

Since the vector of the magnetic field has only an α-component, the electric field has only an r- and ϑ-component (see (5.10)), i.e.,

$$\underline{E}_\alpha = 0$$

and

$$\vec{\underline{E}} = \frac{Z_0}{j \cdot k_0} \cdot \nabla \times \left(\underline{H}_\alpha(r, \vartheta) \cdot \vec{e}_\alpha \right) \tag{5.28}$$

If (5.10) is applied analogously to the magnetic field vector, we have with (5.19)

$$\vec{\underline{E}} = \frac{Z_0}{j \cdot k_0} \cdot \left\{ \frac{1}{r \cdot \sin \vartheta} \cdot \left[\frac{\partial}{\partial \vartheta} \left(\underline{H}_\alpha(r, \vartheta) \cdot \sin \vartheta \right) \right] \right\} \cdot \vec{e}_r$$
$$+ \frac{Z_0}{j \cdot k_0} \cdot \left\{ \frac{1}{r} \cdot \left[-\frac{\partial}{\partial r} \left(r \cdot \underline{H}_\alpha(r, \vartheta) \right) \right] \right\} \cdot \vec{e}_\vartheta$$

$$\vec{\underline{E}} = \frac{Z_0}{j \cdot k_0} \cdot \left\{ \frac{1}{r \cdot \sin \vartheta} \cdot \left[\frac{\partial}{\partial \vartheta} \left(j \cdot \underline{H}_0 \cdot \sin \vartheta \cdot \frac{e^{-j \cdot k_0 \cdot r}}{k_0 \cdot r} \cdot \left[1 + \frac{1}{j \cdot k_0 \cdot r} \right] \cdot \sin \vartheta \right) \right] \right\} \cdot \vec{e}_r$$
$$+ \frac{Z_0}{j \cdot k_0} \left\{ \frac{1}{r} \cdot \left[-\frac{\partial}{\partial r} \left(r \cdot j \cdot \underline{H}_0 \cdot \sin \vartheta \cdot \frac{e^{-j \cdot k_0 \cdot r}}{k_0 \cdot r} \cdot \left[1 + \frac{1}{j \cdot k_0 \cdot r} \right] \right) \right] \right\} \cdot \vec{e}_\vartheta$$

$$\vec{\underline{E}} = \frac{Z_0 \cdot \underline{H}_0}{k_0} \cdot \left\{ \frac{1}{r \cdot \sin \vartheta} \cdot \frac{e^{-j \cdot k_0 \cdot r}}{k_0 \cdot r} \cdot \left[1 + \frac{1}{j \cdot k_0 \cdot r} \right] \cdot \frac{\partial}{\partial \vartheta} \left[(\sin \vartheta)^2 \right] \right\} \cdot \vec{e}_r$$
$$- \frac{Z_0 \cdot \underline{H}_0}{k_0} \cdot \left\{ \frac{\sin \vartheta}{r \cdot k_0} \cdot \left[\frac{\partial}{\partial r} \cdot \left(e^{-j \cdot k_0 \cdot r} \cdot \left[1 + \frac{1}{j \cdot k_0 \cdot r} \right] \right) \right] \right\} \cdot \vec{e}_\vartheta$$

---

[3]Why $\sqrt{\frac{\mu_0}{\varepsilon_0}}$ is referred to as field impedance is explained in Sect. 5.1.2.

$$\vec{E} = \frac{Z_0 \cdot H_0}{k_0} \cdot \left\{ \frac{1}{r \cdot \sin \vartheta} \cdot \frac{e^{-j \cdot k_0 \cdot r}}{k_0 \cdot r} \cdot \left[1 + \frac{1}{j \cdot k_0 \cdot r} \right] \cdot 2 \cdot \sin \vartheta \cdot \cos \vartheta \right\} \cdot \vec{e}_r$$
$$- j \frac{Z_0 \cdot H_0}{k_0} \cdot \left\{ \frac{\sin \vartheta}{r \cdot k_0} \cdot \left(e^{-j \cdot k_0 \cdot r} \cdot (-j \cdot k_0) \cdot \left[1 + \frac{1}{j \cdot k_0 \cdot r} \right] + \left(e^{-j \cdot k_0 \cdot r} \cdot \left[\frac{-1}{j \cdot k_0 \cdot r^2} \right] \right) \right) \right\} \cdot \vec{e}_\vartheta$$

$$\vec{E} = \frac{Z_0 \cdot H_0}{k_0} \cdot \left\{ \frac{1}{r} \cdot \frac{e^{-j \cdot k_0 \cdot r}}{k_0 \cdot r} \cdot \left[1 + \frac{1}{j \cdot k_0 \cdot r} \right] \cdot 2 \cdot \cos \vartheta \right\} \cdot \vec{e}_r$$
$$+ \frac{Z_0 \cdot H_0}{k_0} \cdot \left\{ \frac{\sin \vartheta}{r \cdot k_0} \cdot e^{-j \cdot k_0 \cdot r} \cdot (j \cdot k_0) \cdot \left[1 + \frac{1}{j \cdot k_0 \cdot r} \right] + \left(e^{-j \cdot k_0 \cdot r} \cdot \left[\frac{1}{j \cdot k_0 \cdot r^2} \right] \right) \right\} \cdot \vec{e}_\vartheta$$

$$\vec{E} = \frac{Z_0 \cdot H_0}{k_0} \cdot \left\{ \frac{2 \cdot \cos \vartheta}{r} \cdot \frac{e^{-j \cdot k_0 \cdot r}}{k_0 \cdot r} \cdot \left(1 + \frac{1}{j \cdot k_0 \cdot r} \right) \right\} \cdot \vec{e}_r$$
$$+ \frac{Z_0 \cdot H_0}{k_0} \cdot \left\{ \frac{\sin \vartheta}{r \cdot k_0} \cdot e^{-j \cdot k_0 \cdot r} \cdot \left((j \cdot k_0) + \frac{1}{r} + \frac{1}{j \cdot k_0 \cdot r^2} \right) \right\} \cdot \vec{e}_\vartheta$$

$$\vec{E} = \frac{Z_0 \cdot H_0}{k_0 \cdot r} \cdot \left\{ \frac{2 \cdot \cos \vartheta}{1} \cdot \frac{e^{-j \cdot k_0 \cdot r}}{1} \cdot \left(\frac{1}{k_0 \cdot r} + \frac{1}{j \cdot (k_0 \cdot r)^2} \right) \right\} \cdot \vec{e}_r$$
$$+ \frac{Z_0 \cdot H_0}{k_0 \cdot r} \cdot \left\{ \sin \vartheta \cdot e^{-j \cdot k_0 \cdot r} \cdot \left(j + \frac{1}{r \cdot k_0} + \frac{1}{j \cdot (k_0 \cdot r)^2} \right) \right\} \cdot \vec{e}_\vartheta$$

$$\vec{E} = \left\{ j \cdot \frac{2 \cdot Z_0 \cdot H_0}{k_0 \cdot r} \cdot \cos \vartheta \cdot e^{-j \cdot k_0 \cdot r} \cdot \left(\frac{1}{j \cdot k_0 \cdot r} + \frac{1}{(j \cdot k_0 \cdot r)^2} \right) \right\} \cdot \vec{e}_r$$
$$+ \left\{ j \cdot \frac{Z_0 \cdot H_0}{k_0 \cdot r} \cdot \sin \vartheta \cdot e^{-j \cdot k_0 \cdot r} \cdot \left(1 + \frac{1}{j \cdot r \cdot k_0} + \frac{1}{(j \cdot k_0 \cdot r)^2} \right) \right\} \cdot \vec{e}_\vartheta \tag{5.29}$$

Summary of the results:

$$\underline{H}_\alpha(r, \vartheta) = j \cdot \underline{H}_0 \cdot \sin \vartheta \cdot \frac{e^{-j \cdot k_0 \cdot r}}{k_0 \cdot r} \cdot \left[1 + \frac{1}{j \cdot k_0 \cdot r} \right]$$

$$\underline{E}_r(r, \vartheta) = j \cdot 2 \cdot Z_0 \cdot \underline{H}_0 \cdot \cos \vartheta \cdot \frac{e^{-j \cdot k_0 \cdot r}}{k_0 \cdot r} \cdot \left(\frac{1}{j \cdot k_0 \cdot r} + \frac{1}{(j \cdot k_0 \cdot r)^2} \right) \tag{5.30}$$

$$\underline{E}_\vartheta(r, \vartheta) = j \cdot Z_0 \cdot \underline{H}_0 \cdot \sin \vartheta \cdot \frac{e^{-j \cdot k_0 \cdot r}}{k_0 \cdot r} \cdot \left(1 + \frac{1}{j \cdot r \cdot k_0} + \frac{1}{(j \cdot k_0 \cdot r)^2} \right)$$

5.1.1   Near Field of the Hertzian Dipole

The space in the immediate vicinity of the Hertzian dipole, i.e., the space where $r \ll \lambda$, is called near field of the Hertzian dipole. In this area

$$1 \ll \frac{1}{k_0 \cdot r} = \frac{\lambda}{2 \cdot \pi \cdot r}$$

and

$$\frac{1}{r \cdot k_0} \ll \frac{1}{(k_0 \cdot r)^2}$$

Near-field approximations

$$\underline{H}_\alpha(r, \vartheta)_{\text{near}} = j \cdot \underline{H}_0 \cdot \sin \vartheta \cdot \frac{e^{-j \cdot k_0 \cdot r}}{k_0 \cdot r} \cdot \left(\frac{1}{j \cdot k_0 \cdot r} \right)$$

$$\underline{E}_r(r, \vartheta)_{\text{near}} = j \cdot \frac{2 \cdot Z_0 \cdot \underline{H}_0}{k_0 \cdot r} \cdot \cos \vartheta \cdot e^{-j \cdot k_0 \cdot r} \cdot \left(\frac{1}{(j \cdot k_0 \cdot r)^2} \right) \qquad (5.31)$$

$$\underline{E}_\vartheta(r, \vartheta)_{\text{near}} = j \cdot \frac{Z_0 \cdot \underline{E}_0}{k_0 \cdot r} \cdot \sin \vartheta \cdot e^{-j \cdot k_0 \cdot r} \cdot \left(\frac{1}{(j \cdot k_0 \cdot r)^2} \right)$$

or

$$\underline{H}_\alpha(r, \vartheta)_{\text{near}} = \underline{H}_0 \cdot \sin \vartheta \cdot e^{-j \cdot k_0 \cdot r} \cdot \frac{1}{(k_0 \cdot r)^2}$$

$$\underline{E}_r(r, \vartheta)_{\text{near}} = -j \cdot 2 \cdot Z_0 \cdot \underline{H}_0 \cdot \cos \vartheta \cdot e^{-j \cdot k_0 \cdot r} \cdot \frac{1}{(k_0 \cdot r)^3} \qquad (5.32)$$

$$\underline{E}_\vartheta(r, \vartheta)_{\text{near}} = -j \cdot Z_0 \cdot \underline{H}_0 \cdot \sin \vartheta \cdot e^{-j \cdot k_0 \cdot r} \cdot \frac{1}{(k_0 \cdot r)^3}$$

Also since $k_0 \cdot r \ll 1$

$$e^{-j \cdot k_0 \cdot r} \approx 1$$

Therefore, (5.32) finally can be written as

$$\underline{H}_\alpha(r, \vartheta)_{\text{near}} = \underline{H}_0 \cdot \sin \vartheta \cdot \frac{1}{(k_0 \cdot r)^2}$$

$$\underline{E}_r(r, \vartheta)_{\text{near}} = -j \cdot 2 \cdot Z_0 \cdot \underline{H}_0 \cdot \cos \vartheta \cdot \frac{1}{(k_0 \cdot r)^3} \qquad (5.33)$$

$$\underline{E}_\vartheta(r, \vartheta)_{\text{near}} = -j \cdot Z_0 \cdot \underline{H}_0 \cdot \sin \vartheta \cdot \frac{1}{(k_0 \cdot r)^3}$$

5.1.2 Far Field of the Hertzian Dipole

The area of the far field is defined by

$$1 \gg \frac{\lambda}{2 \cdot \pi \cdot r} = \frac{1}{k_0 \cdot r}$$

or

$$\frac{1}{r \cdot k_0} \gg \frac{1}{(k_0 \cdot r)^2}$$

Far-field relationships

$$\underline{H}_\alpha(r, \vartheta)_{\text{far}} = j \cdot \underline{H}_0 \cdot \sin \vartheta \cdot \frac{e^{-j \cdot k_0 \cdot r}}{k_0 \cdot r}$$

$$\underline{E}_r(r, \vartheta)_{\text{far}} = j \cdot 2 \cdot Z_0 \cdot \underline{H}_0 \cdot \cos \vartheta \cdot \frac{e^{-j \cdot k_0 \cdot r}}{k_0 \cdot r} \cdot \frac{1}{j \cdot k_0 \cdot r} \qquad (5.34)$$

$$\underline{E}_\vartheta(r, \vartheta)_{\text{far}} = j \cdot Z_0 \cdot \underline{H}_0 \cdot \sin \vartheta \cdot \frac{e^{-j \cdot k_0 \cdot r}}{k_0 \cdot r} \cdot$$

or

$$\underline{H}_\alpha(r, \vartheta)_{\text{far}} = j \cdot \underline{H}_0 \cdot \sin \vartheta \cdot \frac{e^{-j \cdot k_0 \cdot r}}{k_0 \cdot r}$$

$$\underline{E}_r(r, \vartheta)_{\text{far}} = 2 \cdot Z_0 \cdot \underline{H}_0 \cdot \cos \vartheta \cdot \frac{e^{-j \cdot k_0 \cdot r}}{(k_0 \cdot r)^2} \qquad (5.35)$$

$$\underline{E}_\vartheta(r, \vartheta)_{\text{far}} = j \cdot Z_0 \cdot \underline{H}_0 \cdot \sin \vartheta \cdot \frac{e^{-j \cdot k_0 \cdot r}}{k_0 \cdot r} \cdot$$

With increasing distance r, the component $\underline{E}_r$ decreases rapidly, so only the two components $\vec{\underline{H}}_\alpha$ and $\vec{\underline{E}}_\vartheta$ are of importance in the far field

$$\vec{\underline{H}}_\alpha(r, \vartheta)_{\text{far}} = \left(j \cdot \underline{H}_0 \cdot \sin \vartheta \cdot \frac{e^{-j \cdot k_0 \cdot r}}{k_0 \cdot r} \right) \cdot \vec{e}_\alpha$$

$$\vec{\underline{E}}_\vartheta(r, \vartheta)_{\text{far}} = \left(j \cdot Z_0 \cdot \underline{H}_0 \cdot \sin \vartheta \cdot \frac{e^{-j \cdot k_0 \cdot r}}{k_0 \cdot r} \right) \cdot \vec{e}_\vartheta \qquad (5.36)$$

The transition between near field and far field is approximately at the so-called boundary radius

$$k_0 \cdot r = k_0 \cdot r_g = 1$$

or

$$r_g = \frac{1}{k_0} \qquad (5.37)$$

At a distance $r = r_g$, near-field and far-field components are in the same order of magnitude. From (5.36), another important result is obtained for the far field

$$\underline{E}_\vartheta = Z_0 \cdot \underline{H}_\alpha \qquad (5.38)$$

or generally

$$\left|\vec{E}\right| = Z_0 \cdot \left|\vec{H}\right| \qquad (5.39)$$

From this relationship, we see why Z_0 is called wave impedance of the electromagnetic field. Equation (5.39) is the analogon to Ohm's law

$$V = R \cdot I$$

Due to the relationship (5.39), it does not matter whether the electric or the magnetic field will be measured.

5.1.3   Electric and Magnetic Field in the Time Domain

In order to transform the electric and magnetic field, presented in (5.29), from the frequency domain to the time domain, Eqs. (5.30) have to be multiplied by $e^{j\omega t}$ according to (4.43) and then the real part has to be taken

$$\vec{E}(r,\vartheta,t) = \mathrm{Re}\left\{\underline{\vec{E}}(r,\vartheta,t)\right\} = \mathrm{Re}\left\{\vec{E}(r,\vartheta) \cdot e^{j \cdot \omega \cdot t}\right\}$$

and

$$\vec{H}(r,\vartheta,t) = \mathrm{Re}\left\{\underline{\vec{H}}(r,\vartheta,t)\right\} = \mathrm{Re}\left\{\vec{H}(r,\vartheta) \cdot e^{j \cdot \omega \cdot t}\right\}$$

The phase angle φ_0 can be set to zero without limiting the general validity.

5.1.3.1 The Component of the Magnetic Field
The component H_α of the magnetic field is

$$H_\alpha(r,\vartheta,t) = \mathrm{Re}\left\{H_\alpha(r,\vartheta) \cdot e^{j \cdot \omega \cdot t}\right\} \qquad (5.40)$$

With $H_\alpha(r,\vartheta)$ from (5.30)[4]

---

[4]Corresponding to (5.16) $\underline{H}_0(t) = \pi \cdot \underline{I} \cdot l/\lambda^2$.

$$H_\alpha(r, \vartheta, t) = \mathrm{Re}\left\{ \left[j \cdot H_0 \cdot \sin\vartheta \cdot \frac{e^{-j \cdot k_0 \cdot r}}{k_0 \cdot r} \cdot \left(1 + \frac{1}{j \cdot k_0 \cdot r} \right) \right] \cdot e^{j \cdot \omega \cdot t} \right\}$$

$$H_\alpha(r, \vartheta, t) = \mathrm{Re}\left\{ j \cdot H_0 \cdot \sin\vartheta \cdot \frac{e^{j(\omega \cdot t - k_0 \cdot r)}}{k_0 \cdot r} \cdot \left(1 + \frac{1}{j \cdot k_0 \cdot r} \right) \right\}$$

$$H_\alpha(r, \vartheta, t) =$$
$$\mathrm{Re}\left\{ j \cdot H_0 \cdot \sin\vartheta \cdot \frac{\cos(\omega \cdot t - k_0 \cdot r) + j \cdot \sin(\omega \cdot t - k_0 \cdot r)}{k_0 \cdot r} \cdot \left(1 + \frac{1}{j \cdot k_0 \cdot r} \right) \right\}$$

$$H_\alpha(r, \vartheta, t) =$$
$$\frac{H_0 \cdot \sin\vartheta}{k_0 \cdot r} \cdot \mathrm{Re}\left\{ j \cdot \left[\cos(\omega \cdot t - k_0 \cdot r) + j \cdot \sin(\omega \cdot t - k_0 \cdot r) \right] \cdot \left(1 + \frac{1}{j \cdot k_0 \cdot r} \right) \right\}$$

$$H_\alpha(r, \vartheta, t) = \frac{H_0 \cdot \sin\vartheta}{k_0 \cdot r} \cdot \mathrm{Re}\left\{ \left[j \cdot \cos(\omega \cdot t - k_0 \cdot r) - \sin(\omega \cdot t - k_0 \cdot r) \right] \cdot \left(1 + \frac{1}{j \cdot k_0 \cdot r} \right) \right\}$$

$$H_\alpha(r, \vartheta, t) = \frac{H_0 \cdot \sin\vartheta}{k_0 \cdot r} \cdot \left[\frac{\cos(\omega \cdot t - k_0 \cdot r)}{k_0 \cdot r} - \sin(\omega \cdot t - k_0 \cdot r) \right] \qquad (5.41)$$

The first summand in the square bracket of (5.41) can be omitted for the far field because

$$1 \gg \frac{1}{k_0 \cdot r}$$

Therefore

$$H_\alpha(r, \vartheta, t)_{\mathrm{far}} = -\frac{H_0 \cdot \sin\vartheta}{k_0 \cdot r} \cdot \left[\sin(\omega \cdot t - k_0 \cdot r) \right] \qquad (5.42)$$

5.1.3.2 Electric Field Component in r-Direction

The component E_r of the electric field is

$$E_r(r, \vartheta, t) = \mathrm{Re}\left\{ E_r(r, \vartheta) \cdot e^{j \cdot \omega \cdot t} \right\} \qquad (5.43)$$

With (5.30)

$$E_r(r, \vartheta, t) = \mathrm{Re}\left\{ \left[j \cdot 2 \cdot Z_0 \cdot H_0 \cdot \cos\vartheta \cdot \frac{e^{-j \cdot k_0 \cdot r}}{k_0 \cdot r} \cdot \left(\frac{1}{j \cdot k_0 \cdot r} + \frac{1}{(j \cdot k_0 \cdot r)^2} \right) \right] \cdot e^{j \cdot \omega \cdot t} \right\}$$

$$E_r(r, \vartheta, t) = \frac{2 \cdot Z_0 \cdot H_0 \cdot \cos\vartheta}{k_0 \cdot r} \cdot \mathrm{Re}\left\{ j \cdot e^{j \cdot (\omega \cdot t - k_0 \cdot r)} \cdot \left(\frac{1}{j \cdot k_0 \cdot r} + \frac{1}{(j \cdot k_0 \cdot r)^2} \right) \right\}$$

$$E_r(r, \vartheta, t) = \frac{2 \cdot Z_0 \cdot H_0 \cdot \cos\vartheta}{k_0 \cdot r} \cdot$$
$$\mathrm{Re}\left\{ j \cdot \left[\cos(\omega \cdot t - k_0 \cdot r) + j \cdot \sin(\omega \cdot t - k_0 \cdot r) \right] \cdot \left(\frac{1}{j \cdot k_0 \cdot r} + \frac{1}{(j \cdot k_0 \cdot r)^2} \right) \right\}$$

$$E_r(r,\vartheta,t) = \frac{2 \cdot Z_0 \cdot H_0 \cdot \cos\vartheta}{k_0 \cdot r} \cdot$$

$$\text{Re}\left\{ \left[j \cdot \cos\left(\omega \cdot t - k_0 \cdot r\right) - \sin\left(\omega \cdot t - k_0 \cdot r\right) \right] \cdot \left(\frac{1}{j \cdot k_0 \cdot r} + \frac{1}{(j \cdot k_0 \cdot r)^2} \right) \right\}$$

$$E_r(r,\vartheta,t) = \frac{2 \cdot Z_0 \cdot H_0 \cdot \cos\vartheta}{(k_0 \cdot r)^2} \cdot \left(\cos\left(\omega \cdot t - k_0 \cdot r\right) + \frac{\sin\left(\omega \cdot t - k_0 \cdot r\right)}{k_0 \cdot r} \right) \quad (5.44)$$

In the far field, the second summand in the square bracket of (5.44) is to be omitted

$$E_r(r,\vartheta,t)_{\text{far}} = \frac{2 \cdot Z_0 \cdot H_0 \cdot \cos\vartheta}{(k_0 \cdot r)^2} \cdot \cos\left(\omega \cdot t - k_0 \cdot r\right) \quad (5.45)$$

5.1.3.3 Electrical Field Component in ϑ-Direction

The component E_ϑ of the electric field is

$$E_\vartheta(r,\vartheta,t) = \text{Re}\left\{ E(r,\vartheta) \cdot e^{j\cdot\omega\cdot t} \right\} \quad (5.46)$$

With (5.30)

$$E_\vartheta(r,\vartheta,t) = \text{Re}\left\{ \left[j \cdot Z_0 \cdot H_0 \cdot \sin\vartheta \cdot \frac{e^{-j\cdot k_0 \cdot r}}{k_0 \cdot r} \cdot \left(1 + \frac{1}{j \cdot r \cdot k_0} + \frac{1}{(j \cdot k_0 \cdot r)^2} \right) \right] \cdot e^{j\cdot\omega\cdot t} \right\}$$

$$E_\vartheta(r,\vartheta,t) = \frac{Z_0 \cdot H_0 \cdot \sin\vartheta}{k_0 \cdot r} \text{Re}\left\{ j \cdot e^{j\cdot(\omega\cdot t - k_0 \cdot r)} \cdot \left(1 + \frac{1}{j \cdot r \cdot k_0} + \frac{1}{(j \cdot k_0 \cdot r)^2} \right) \right\}$$

$$E_\vartheta(r,\vartheta,t) = \frac{Z_0 \cdot H_0 \cdot \sin\vartheta}{k_0 \cdot r} \cdot$$

$$\text{Re}\left\{ \left[j \cdot \cos\left(\omega \cdot t - k_0 \cdot r\right) - \sin\left(\omega \cdot t - k_0 \cdot r\right) \right] \cdot \left(1 + \frac{1}{j \cdot r \cdot k_0} - \frac{1}{(k_0 \cdot r)^2} \right) \right\}$$

$$E_\vartheta(r,\vartheta,t) = \frac{Z_0 \cdot H_0 \cdot \sin\vartheta}{k_0 \cdot r} \cdot$$
$$\left[\frac{\cos\left(\omega \cdot t - k_0 \cdot r\right)}{r \cdot k_0} - \sin\left(\omega \cdot t - k_0 \cdot r\right) \cdot \left(1 - \frac{1}{(k_0 \cdot r)^2} \right) \right] \quad (5.47)$$

The first summand in the square bracket and the summand

$$\frac{1}{(k_0 \cdot r)^2}$$

in the round bracket are negligible in the far field. Therefore

$$E_\vartheta(r,\vartheta,t)_{\text{far}} = -\frac{Z_0 \cdot H_0 \cdot \sin\vartheta}{k_0 \cdot r} \cdot \sin\left(\omega \cdot t - k_0 \cdot r\right) \quad (5.48)$$

Summary of the results:

$$H_\alpha(r, \vartheta, t) = \frac{H_0 \cdot \sin \vartheta}{k_0 \cdot r} \cdot \left[\frac{\cos(\omega \cdot t - k_0 \cdot r)}{k_0 \cdot r} - \sin(\omega \cdot t - k_0 \cdot r) \right]$$

$$E_r(r, \vartheta, t) = \frac{2 \cdot Z_0 \cdot H_0 \cdot \cos \vartheta}{(k_0 \cdot r)^2} \cdot \left(\cos(\omega \cdot t - k_0 \cdot r) + \frac{\sin(\omega \cdot t - k_0 \cdot r)}{k_0 \cdot r} \right)$$

$$E_\vartheta(r, \vartheta, t) = \frac{Z_0 \cdot H_0 \cdot \sin \vartheta}{k_0 \cdot r} \cdot$$

$$\left[\frac{\cos(\omega \cdot t - k_0 \cdot r)}{r \cdot k_0} - \sin(\omega \cdot t - k_0 \cdot r) \cdot \left(1 - \frac{1}{(k_0 \cdot r)^2} \right) \right] \tag{5.49}$$

Near field ($k_0 \cdot r \ll 1$):

$$H_\alpha(r, \vartheta, t)_{\text{near}} = \frac{H_0 \cdot \sin \vartheta}{(k_0 \cdot r)^2} \cdot \cos(\omega \cdot t - k_0 \cdot r)$$

$$E_r(r, \vartheta, t)_{\text{near}} = \frac{2 \cdot Z_0 \cdot H_0 \cdot \cos \vartheta}{(k_0 \cdot r)^3} \cdot \sin(\omega \cdot t - k_0 \cdot r) \tag{5.50}$$

$$E_\vartheta(r, \vartheta, t)_{\text{near}} = \frac{Z_0 \cdot H_0 \cdot \sin \vartheta}{(k_0 \cdot r)^3} \cdot \sin(\omega \cdot t - k_0 \cdot r)$$

Far field $\left(1 \gg \frac{1}{k_0 \cdot r} \text{ and } \frac{1}{r \cdot k_0} \gg \frac{1}{(k_0 \cdot r)^2} \right)$:

$$H_\alpha(r, \vartheta, t)_{\text{far}} = - \frac{H_0 \cdot \sin \vartheta}{k_0 \cdot r} \cdot \sin(\omega \cdot t - k_0 \cdot r)$$

$$E_r(r, \vartheta, t)_{\text{far}} = \frac{2 \cdot H_0 \cdot \hat{H}_0 \cdot \cos \vartheta}{(k_0 \cdot r)^2} \cdot \cos(\omega \cdot t - k_0 \cdot r) \tag{5.51}$$

$$E_\vartheta(r, \vartheta, t)_{\text{far}} = - \frac{Z_0 \cdot H_0 \cdot \sin \vartheta}{k_0 \cdot r} \cdot \sin(\omega \cdot t - k_0 \cdot r)$$

5.1.4 Energy Flux of the Hertzian Dipole

5.1.4.1 Energy Flux of the Hertzian Dipole in the Near Field

From (5.50) it can be seen that in the near field the magnetic field has a phase shift of 90° or $\pi/2$ compared to the electric field components. ($\cos \alpha = \sin(90° \pm \alpha) = \sin(\pi/2 \pm \alpha)$). According to (4.112), the Poynting vector, and thus the energy flux in the electromagnetic field, is the vector product of the electric and the magnetic field

$$\vec{S} = \vec{E} \times \vec{H}$$

The energy flux in the r-direction of the near field is defined only by $H_\alpha(r, \vartheta, t)$ and $E_\vartheta(r, \vartheta, t)$ (see Fig. 5.4). Both components are perpendicular to each other; consequently, the magnitude of the Poynting vector is

$$\left|\vec{S}\right| = S = E_{\vartheta/\text{near}} \cdot H_{\alpha/\text{near}} \cdot \sin 90° = E_{\vartheta/\text{near}} \cdot H_{\alpha/\text{near}} \tag{5.52}$$

Figure 5.5 shows the time functions of both the electric field component $E_\vartheta(t)$ and the magnetic field component $H_\alpha(t)$ for time-harmonic dependence according to (5.50) for fixed values of r and ϑ. Also, the product $S(t) = E_\vartheta(t) \cdot H_\alpha(t)$ is shown in Fig. 5.5.

It can be seen that within half a period the energy flux $S(t) = E_\vartheta(t) \cdot H_\alpha(t)$ is half positive and half negative. This means that for half a period the energy flux is directed

Fig. 5.4 Field components of the Hertzian dipole

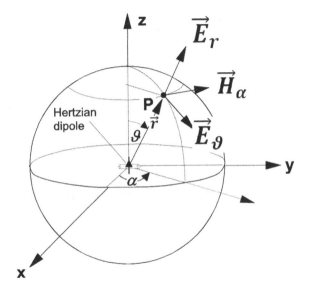

Fig. 5.5 Reactive power

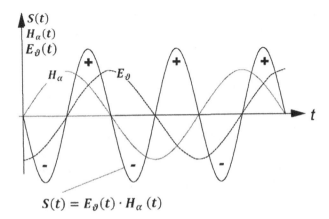

$$S(t) = E_\vartheta(t) \cdot H_\alpha(t)$$

into a positive r-direction away from the Hertzian dipole. In the next half period, the energy flows back into negative r-direction toward the dipole. In the near field, the energy flux is a reactive power, i.e., a power that oscillates between the dipole and the surrounding space. Consequently, the field components of the near field, according to (5.50), do not contribute to the power radiation of the Hertzian dipole.

5.1.4.2 Energy Flux of the Hertzian Dipole in the Far Field

According to (5.51), the components $H_\alpha(r, \vartheta, t)$ and $E_\vartheta(r, \vartheta, t)$ in the far field are

$$
\begin{aligned}
H_\alpha(r, \vartheta, t)_{\text{far}} &= - \frac{H_0 \cdot \sin \vartheta}{k_0 \cdot r} \cdot \sin(\omega \cdot t - k_0 \cdot r) \\
E_\vartheta(r, \vartheta, t)_{\text{far}} &= - \frac{Z_0 \cdot H_0 \cdot \sin \vartheta}{k_0 \cdot r} \cdot \sin(\omega \cdot t - k_0 \cdot r)
\end{aligned}
\tag{5.53}
$$

With the Poynting vector (4.112)

$$
\vec{S} = \vec{E} \times \vec{H} = E_{\vartheta/\text{far}} \cdot \vec{e}_\vartheta \times H_{\alpha/\text{far}} \cdot \vec{e}_\alpha = \vec{E}_{\vartheta/\text{far}} \times \vec{H}_{\alpha/\text{far}}
\tag{5.54}
$$

The vectors $\vec{S}$, $\vec{E}_{\vartheta/\text{far}}$, and $\vec{H}_{\alpha/\text{far}}$ form a right-hand screw. Since the electric field vector $\vec{E}_{\vartheta/\text{far}}$ and the magnetic field vector $\vec{H}_{\alpha/\text{far}}$ have the same phase angle in the far field, the Poynting vector is always orientated into the positive r-direction, i.e., energy is transported away from the Hertzian dipole into the surrounding space.

5.1.5 Field Lines of the Hertzian Dipole

With (5.51), the field lines of the Hertzian dipole can be computed for the far field as a function of time t. On this basis, the electromagnetic waves emanating from the Hertzian dipole can be illustrated.

Field lines are imaginary lines that illustrate the directions of the vectors in vector fields. At each point of a field line, the tangent to the field line matches the direction of the vector in that field point. The current density lines shown in Fig. 1.4 are field lines. The corresponding vectors of the field are depicted in Fig. 1.6. Figure 2.13 shows the electric field lines of an electric dipole. Magnetic field lines can be made visible with the test arrangement sketched in Fig. 3.2 and 3.3. The density of field lines reflects the intensity of the field.

From (5.49), it can be seen that the magnetic field of the Hertzian dipole has only a component in α-direction. Consequently, as could be expected from the orientation of the dipole in the z-direction, the magnetic field lines are concentric circles around the z-axis. The electric field on, the other hand, has both a component in r-direction and a component in ϑ-direction. They are both independent of the angle α. Consequently, the image of the electric field lines is rotationally symmetric to the z-axis. It is therefore sufficient if only the surface $\alpha = $ const is considered.

Starting point for the calculation of an electric field line of the Hertzian dipole, at a certain time t_1, and for a certain value of the wave number $k_0 = 2 \cdot \pi / \lambda$, or a certain frequency $\omega = 2 \cdot \pi \cdot f$, is an appropriately chosen point r_0, ϑ_0, in a surface $\alpha = \text{const}$, of the space. For this point, the vectorial sum

$$\vec{E} = E_r(r_0, \vartheta_0) \cdot \vec{e}_r + E_\vartheta(r_0, \vartheta_0) \cdot \vec{e}_\vartheta$$

is calculated and from the result the direction of the vector of the electric field is determined. The next point of the electric field line is obtained by progressing into the direction of this vector by an infinitesimal step, i.e., in the direction of the tangent of the field line. This gives the second point r_1, ϑ_1 of the electric field line. For this second point, the components $E_r(r_1, \vartheta_1)$ and $E_\vartheta(r_1, \vartheta_1)$ and their vectorial sum are calculated. To obtain the third point of the electric field line, one must again advance by an infinitesimal distance into the direction of the field strength vector $\vec{E} = E_r(r_1, \vartheta_1) \cdot \vec{e}_r + E_\vartheta(r_1, \vartheta_1) \cdot \vec{e}_\vartheta$, etc.

In Fig. 5.6, a field line for the starting point $r_0 = 1{,}64$ and $\vartheta_0 = 90°$ and for a frequency 500 MHz at time $t = T = 1/f$ is depicted. The field line was created with a small program of the mathematics software Mathcad.

Figure 5.6 shows that this electric field line of the Hertzian dipole is closed in contrast to a field line of the electrostatic field. In the immediate vicinity of the Hertzian dipole, the field lines originate and end in the Hertzian dipole. They move away from the dipole as time goes on. After half a period, the current direction in the dipole changes and the field lines in the immediate vicinity of the dipole also change this direction. The more

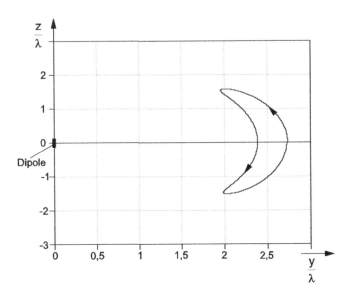

Fig. 5.6 Electrical field line of the Hertzian dipole in the surface $\alpha = 90°$: Starting point: $y/\lambda = 2{,}735, z/\lambda = 0, f = 500$ MHz, $t = 2$ ns, step width: 0,00179

Fig. 5.7 Electrical field lines
of the Hertzian dipole, surface
$\alpha = $ const, range $0 < r/\lambda < 2$

distant field lines "notice" the change of the current direction, due to the finite velocity of propagation, with delay and therefore detach themselves from the dipole.

Figure 5.7 shows the electric field lines of the Hertzian dipole in the surface $\alpha = $ const in the range $0 < r/\lambda < 2$ for one point in time. Addresses of four websites where animations of the wave propagation of the Hertzian dipole can be viewed are listed in the literature reference.

5.1.6   Radiation Pattern of the Hertzian Dipole

The radiation pattern of an antenna describes the directional dependence of the energy flux density, i.e., the power per unit area, in the far field of the dipole. According to (4.124), the energy density, transported in the electromagnetic field into the direction of the Poynting vector in the case of time-harmonic, is

$$\vec{S} = \frac{1}{2} \cdot \mathrm{Re}\left\{\underline{\vec{E}} \times \underline{\vec{H}}^{*}\right\} \tag{5.55}$$

In the far field, only the E_{ϑ} components of the electric field and the H_{α} component of the magnetic field are responsible for the power transport into r-direction. According to (5.36), these field components are in phase. Therefore, the power flux density in (5.55) is an active power density. Thus

$$\vec{S}_{\text{active}} = \frac{1}{2} \cdot \mathrm{Re}\left\{\left(\underline{\vec{E}}_{\vartheta}(r, \vartheta)_{\text{far}} \cdot \underline{\vec{H}}_{\alpha}(r, \vartheta)^{*}_{\text{far}}\right)\right\}$$

or

$$\vec{S}_{\text{active}} = \frac{1}{2} \cdot \text{Re}\left\{ \left(\underline{E}_\vartheta (r, \vartheta)_{\text{far}} \cdot \underline{H}_\alpha (r, \vartheta)^*_{\text{far}} \right) \cdot (\vec{e}_\vartheta \times \vec{e}_\alpha) \right\} \tag{5.56}$$

With (5.36) and with $j = e^{+j \cdot \pi/2}$

$$\vec{S}_{\text{active}} = \frac{1}{2} \cdot \text{Re}\left\{ \left(Z_0 \cdot \underline{H}_0 \cdot \sin \vartheta \cdot \frac{e^{-j \cdot k_0 \cdot r + j \cdot \pi/2}}{k_0 \cdot r} \right) \cdot \left(\underline{H}_0 \cdot \sin \vartheta \cdot \frac{e^{-j \cdot k_0 \cdot r + j \cdot \pi/2}}{k_0 \cdot r} \right)^* \cdot (\vec{e}_\vartheta \times \vec{e}_\alpha) \right\}$$

$$\vec{S}_{\text{active}} = \frac{1}{2} \cdot \text{Re}\left\{ \left(Z_0 \cdot \underline{H}_0 \cdot \sin \vartheta \cdot \frac{e^{-j \cdot k_0 \cdot r + j \cdot \pi/2}}{k_0 \cdot r} \right) \cdot \left(\underline{H}_0 \cdot \sin \vartheta \cdot \frac{e^{-(-j \cdot k_0 \cdot r + j \cdot \pi/2)}}{k_0 \cdot r} \right) \cdot \vec{e}_r \right\}$$

$$\vec{S}_{\text{active}} = \frac{1}{2} \cdot \frac{Z_0 \cdot H_0^2}{k_0^2 \cdot r^2} \cdot (\sin \vartheta)^2 \cdot \vec{e}_r$$

or with (5.39) (5.57)

$$\vec{S}_{\text{active}} = \frac{1}{2} \cdot \frac{E_0^2}{Z_0 \cdot (k_0^2 \cdot r^2)} \cdot (\sin \vartheta)^2 \cdot \vec{e}_r$$

The power density in the far field of the Hertzian dipole depends, according to (5.57), on the square of the sine of ϑ (cf. Fig. 5.4). The maximum power is radiated into the direction $\vartheta = 90°$. This dependence is represented in the radiation pattern as a function of ϑ in relation to the maximum value, either linearly or logarithmically as relative power level. The radiation pattern $C(\vartheta)$ or $c(\vartheta)$ of the Hertzian dipole is

$$C(\vartheta) = \frac{\frac{1}{2} \cdot \frac{Z_0 \cdot H^2}{k_0^2 \cdot r^2} \cdot (\sin \vartheta)^2}{\frac{1}{2} \cdot \frac{Z_0 \cdot H_0^2}{k_0^2 \cdot r^2}}$$

$$C(\vartheta) = (\sin \vartheta)^2 \tag{5.58}$$

or

$$c(\vartheta) = 10 \cdot \log_{10} (\sin \vartheta)^2 \text{ dB} \tag{5.59}$$

The radiation patterns $C(\vartheta)$ und $c(\vartheta)$ of the Hertzian dipole are shown in Figs. 5.8 and 5.9. No power is radiated into the direction of the dipole axis ($\vartheta = 0°$ and $\vartheta = 180°$).

5.1.7 Radiated Power of the Hertzian Dipole

The total power radiated into the far field of the Hertzian dipole P_{rad} is obtained by integrating the power density $\vec{S}$ over the spherical surface A with the radius r.

$$P_{\text{rad}} = \oint_A \vec{S} \cdot d\vec{A} \tag{5.60}$$

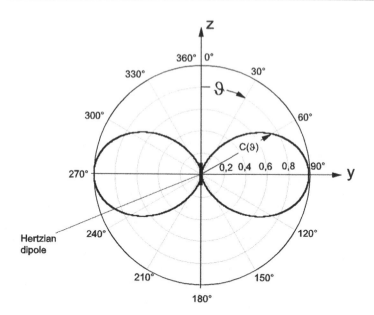

Fig. 5.8 Radiation pattern $C(\vartheta)$ of the Hertzian dipole (linear representation in polar coordinates)

Fig. 5.9 Radiation pattern
$c(\vartheta)$ of the Hertzian dipole
(logarithmic representation in
Cartesian coordinates)

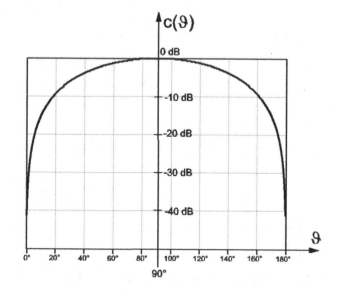

From (5.57)

$$P_{\text{rad}} = \oint_A \frac{1}{2} \cdot \frac{Z_0 \cdot H_0^2}{k_0^2 \cdot r^2} \cdot (\sin \vartheta)^2 \cdot dA \tag{5.61}$$

Based on Fig. 3.27 with the surface element dA

$$dA = (r \cdot d\vartheta) \cdot (r \cdot \sin \vartheta \, d\alpha)$$

With this equation and $(5.61)^5$

$$P_{rad} = \frac{1}{2} \cdot \frac{Z_0 \cdot H_0^2}{k_0^2 \cdot r^2} \cdot \int_{\vartheta=0}^{\pi} \int_{\alpha=0}^{2 \cdot \pi} (\sin \vartheta)^2 \cdot (r \cdot d\vartheta) \cdot (r \cdot \sin \vartheta \, d\alpha)$$

$$P_{rad} = \frac{1}{2} \cdot \frac{Z_0 \cdot H_0^2}{k_0^2} \cdot \int_{\vartheta=0}^{\pi} (\sin \vartheta)^3 \cdot d\vartheta \cdot \int_{\alpha=0}^{2 \cdot \pi} d\alpha$$

$$P_{rad} = \frac{1}{2} \cdot \frac{Z_0 \cdot H_0^2}{k_0^2} \cdot [\alpha]_0^{2 \cdot \pi} \cdot \int_{\vartheta=0}^{\pi} (\sin \vartheta)^3 \cdot d\vartheta$$

$$P_{rad} = \frac{1}{2} \cdot \frac{Z_0 \cdot H_0^2}{k_0^2} \cdot 2 \cdot \pi \cdot \left[-\cos \vartheta + \frac{1}{3} \cdot (\cos \vartheta)^3 \right]_0^{\pi}$$

$$P_{rad} = \frac{1}{2} \cdot \frac{Z_0 \cdot H_0^2}{k_0^2} \cdot 2 \cdot \pi \cdot \left[[(+1) - (-1)] + \frac{1}{3} \cdot ((-1) - (1)) \right]$$

$$P_{rad} = \frac{1}{2} \cdot \frac{Z_0 \cdot H_0^2}{k_0^2} \cdot 2 \cdot \pi \cdot \left[2 - \frac{2}{3} \right]$$

$$P_{rad} = \frac{1}{2} \cdot \frac{Z_0 \cdot H_0^2}{k_0^2} \cdot 2 \cdot \pi \cdot \frac{4}{3} \qquad (5.62)$$

With

$$H_0 = \frac{\pi \cdot I \cdot l}{\lambda^2} \text{ and } k_0 = \frac{2 \cdot \pi}{\lambda}$$

from (5.16) and (5.18), Eq. (5.62) changes into

$$P_{rad} = \frac{1}{2} \cdot Z_0 \cdot \frac{\left(\frac{\pi \cdot l \cdot l}{\lambda^2} \right)^2}{\left(\frac{2 \cdot \pi}{\lambda} \right)^2} \cdot 2 \cdot \pi \cdot \frac{4}{3}$$

Thus, we obtain the power radiated by the Hertzian dipole

$$P_{rad} = \frac{\pi}{3} \cdot Z_0 \cdot \left(\frac{l}{\lambda} \right)^2 \cdot I^2 \qquad (5.63)$$

---

^{5}Integral tables, e.g., in [7].

5.1.8 Radiation Resistance of the Hertzian Dipole

The power P_{rad}, radiated by the Hertzian dipole, is supplied to the dipole by a generator. The Hertzian dipole represents the load to this generator. The dipole could thus be replaced by a resistance, the so-called radiation resistance R_{rad} in which the same power would be converted into heat as radiated by the Hertzian dipole. Therefore

$$P_{rad} = \frac{1}{2} \cdot I^2 \cdot R_{rad} \tag{5.64}$$

$I =$ amplitude of the current, $I/\sqrt{2} =$ RMS current
   With (5.63)

$$\frac{\pi}{3} \cdot Z_0 \cdot \left(\frac{l}{\lambda}\right)^2 \cdot I^2 = \frac{1}{2} \cdot I^2 \cdot R_{rad}$$

or

$$R_{rad} = \frac{2 \cdot \pi}{3} \cdot Z_0 \cdot \left(\frac{l}{\lambda}\right)^2 \tag{5.65}$$

5.1.9 Gain of a Transmitting Antenna

In addition to the radiation pattern of an antenna, its gain is important for the design of radio links. The gain is the ratio of the maximum power density of an antenna to under consideration the power density of an isotropic radiator. The isotropic radiator is a fictitious antenna radiating power evenly into all directions of the space. Thus, the power density $S_{isotrop}$ of the isotropic radiator at a distance of r from the antenna is

$$S_{isotrop} = \frac{P_{rad}}{4 \cdot \pi \cdot r^2} \tag{5.66}$$

The maximum power density of the Hertzian dipole is radiated at $\vartheta = 90°$ (see (5.57)):

$$S_{Hertz/max} = \frac{1}{2} \cdot \frac{Z_0 \cdot H_0^2}{k_0^2 \cdot r^2}$$

With

$$H_0 = \frac{\pi \cdot I \cdot l}{\lambda^2} \text{ und } k_0 = \frac{2 \cdot \pi}{\lambda}$$

$$S_{Hertz/max} = \frac{1}{2} \cdot \frac{Z_0 \cdot \left(\frac{\pi \cdot I \cdot l}{\lambda^2},\right)^2}{\left(\frac{2 \cdot \pi}{\lambda}\right)^2 \cdot r^2}$$

Fig. 5.10 Equivalent circuit diagram of the receiver antenna system

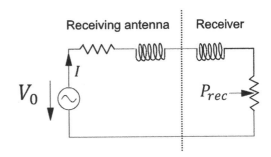

or

$$S_{\text{Hertz/max}} = \frac{1}{8} \cdot Z_0 \frac{I^2 \cdot l^2}{\lambda^2 \cdot r^2} \tag{5.67}$$

From (5.66) and (5.67), the gain factor G_{Hertz} of the Hertzian dipole is

$$G_{\text{Hertz}} = \frac{S_{\text{Hertz/max}}}{S_{\text{isotrop}}} = \frac{1}{8} \cdot Z_0 \frac{I^2 \cdot l^2}{\lambda^2 \cdot r^2} \cdot \frac{4 \cdot \pi \cdot r^2}{P_{\text{rad}}} = \frac{1}{8} \cdot Z_0 \cdot \frac{I^2 \cdot l^2}{\lambda^2 \cdot r^2} \cdot \frac{4 \cdot \pi \cdot r^2}{\frac{\pi}{3} \cdot Z_0 \cdot \left(\frac{I}{\lambda}\right)^2 \cdot l^2}$$

$$G_{\text{Hertz}} = \frac{3}{2} \tag{5.68}$$

The logarithmic gain of the Hertzian dipole is

$$g_{\text{Hertz}} = 10 \cdot \log_{10}\left(\frac{3}{2}\right) \text{ dB} = 1.76 \text{ dB} \tag{5.69}$$

5.1.10 Effective Area[6] of a Receiving Antenna

The task of a receiving antenna is to intercept electrical power from the electromagnetic field and to feed it to the receiver. Figure 5.10 shows the equivalent circuit of the system consisting of the antenna and the receiver.

The receiving antenna can be regarded as a voltage source, where the voltage V_0 is caused by the electric field at the receiving location. The internal complex impedance of the antenna is $R_{\text{rad}} + j \cdot X_A$. The input impedance of the receiver $R_L + j \cdot X_L$ is the load of the antenna. To deliver maximum power from the antenna to the receiver, the antenna must be terminated with an impedance complex conjugate to the antenna impedance $R_{\text{rad}} + j \cdot X_A$, i.e.,

$$R_{\text{rad}} + j \cdot X_A = R_L - j \cdot X_L$$

---

[6]or receiving cross section

and thus

$$R_{\text{rad}} = R_L \text{ and } X_A = -X_L \tag{5.70}$$

With the condition $R_L = R_{\text{rad}}$, the power P_{rec}, delivered to the load resistor R_L of the receiver, is

$$P_{\text{rec}} = \frac{1}{2} \cdot \hat{I}^2 \cdot R_L = \frac{1}{2} \cdot \frac{V_0^2}{(R_{\text{rad}} + R_L)^2} \cdot R_L = \frac{1}{2} \cdot \frac{V_0^2}{(2 \cdot R_{\text{rad}})^2} \cdot R_L = \frac{V_0^2}{8 \cdot R_{\text{rad}}} = \frac{V_0^2}{8 \cdot R_L} \tag{5.71}$$

The factor $1/2$ in (5.71) takes into account that I is the current amplitude, and the power is the product of the square of the effective value of the current and the load resistance (Fig. 5.11).

The voltage V_0 is the source voltage of the receiving antenna. It is generated by the electric field $\vec{E}$ of the incoming wave, acting along the antenna of length l. According to (1.13), a voltage $|V_{\text{ab}}|$ is generated between the endpoints a and b of a conductor

$$|V_{\text{ab}}| = V_0 = \int_a^b \vec{E} \cdot d\vec{s} \tag{5.72}$$

The length l of the Hertzian dipole is small compared to the wavelength of the incoming electromagnetic wave. As a result, the field along the dipole is assumed to be constant. In this case

$$V_0 = \int_a^b \vec{E} \cdot d\vec{s} = E \cdot l \cdot \cos \beta \tag{5.73}$$

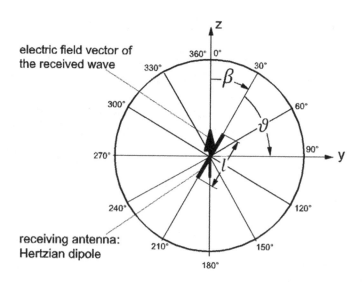

Fig. 5.11 Orientation of the receiving antenna

With (5.71) the receiving power is

$$P_{\text{rec}/\beta} = \frac{V_0^2}{8 \cdot R_L} = \frac{(E \cdot l)^2}{8 \cdot R_L} \cdot (\cos \beta)^2 \tag{5.74}$$

For $\beta = 0$, the received power is maximum. The radiation pattern C_{rec} of the Hertzian dipole, as a receiving antenna, is obtained from (5.74) and (5.71)

$$C_{\text{rec}}(\beta) = \frac{P_{\text{rec}/\beta}}{P_{\text{rec}}} = \frac{\frac{(E \cdot l)^2}{8 \cdot R_L} \cdot (\cos \beta)^2}{\frac{(E \cdot l)^2}{8 \cdot R_L}} = \frac{\frac{(E \cdot l)^2}{8 \cdot R_L} \cdot [\cos (90° - \vartheta)]^2}{\frac{(E \cdot l)^2}{8 \cdot R_L}}$$

or

$$C_{\text{rec}}(\vartheta) = (\sin \vartheta)^2$$

Therefore, the radiation pattern of the Hertzian dipole, as a receiving antenna, corresponds to the radiation pattern of the Hertzian dipole as a transmitting antenna (see (5.58)). This is true in all cases. Radiation pattern, gain, and radiation resistance of an antenna are independent of whether it is used as a transmitting or a receiving antenna. If this were not the case, an exchange of transmitting and receiving antennas, at constant transmitting power, would result in an increase or decrease of the received power, which would contradict the law of conservation of energy.

The received power P_{rec} results from the fact that the receiving antenna is able to extract power from the power density of the electromagnetic field. This ability of the receiving antenna can be characterized by the so-called effective area A_{eff} of the antenna. The received power P_{rec} is then obtained from the product of the effective area A_{eff} and the incoming power density S_{rec} at the location of the receiving antenna

$$P_{\text{rec}} = S_{\text{rec}} \cdot A_{\text{eff}} \tag{5.75}$$

The power density S_{rec} in the far field of the transmitting Hertzian dipole (5.57) is

$$S_{\text{rec}} = \frac{1}{2} \cdot \frac{Z_0 \cdot H_0^2}{k_0^2 \cdot r^2} \cdot (\sin \vartheta)^2 \tag{5.76}$$

With (5.71) and (5.76)

$$A_{\text{eff}} = \frac{P_{\text{rec}}}{S_{\text{rec}}} = \frac{\frac{V_0^2}{8 \cdot R_{\text{rad}}}}{\frac{1}{2} \cdot \frac{Z_0 \cdot H_0^2}{k_0^2 \cdot r^2} \cdot (\sin \vartheta)^2} \tag{5.77}$$

A receiving Hertzian dipole is optimally aligned, when oriented in parallel to the electric field vector $\vec{E}_\vartheta$ of the electromagnetic wave arriving at ($\beta = 0°$) (see (5.73))

$$V_{\text{ab}} = V_0 = \int_a^b \vec{E} \cdot d\vec{s} = E \cdot l \tag{5.78}$$

The definition of the effective area A_{eff} always refers to this case. Corresponding to (5.36)

$$V_0 = E \cdot l = \frac{Z_0 \cdot H_0 \cdot \sin \vartheta}{k_0 \cdot r} \cdot l. \tag{5.79}$$

With this relationship and (5.65), the equation (5.77) changes into

$$A_{\text{eff/Hertz}} = \frac{P_{\text{rec}}}{S_{\text{rec}}} = \frac{\left(\frac{Z_0 \cdot H_0 \cdot \sin \vartheta}{k_0 \cdot r}\right)^2 \frac{l^2}{8 \cdot R_{\text{rad}}}}{\frac{1}{2} \cdot \frac{Z_0 \cdot H_0^2}{k_0^2 \cdot r^2} \cdot (\sin \vartheta)^2} = \frac{\left(\frac{Z_0 \cdot H_0 \cdot \sin \vartheta}{k_0 \cdot r}\right)^2 \cdot \frac{l^2}{8 \cdot \left[\frac{2 \cdot \pi}{3} \cdot Z_0 \cdot \left(\frac{l}{\lambda}\right)^2\right]}}{\frac{1}{2} \cdot \frac{Z_0 \cdot H_0^2}{k_0^2 \cdot r^2} \cdot (\sin \vartheta)^2}$$

Thus the effective area of the Hertzian dipole is

$$A_{\text{eff/Hertz}} = \frac{3 \cdot \lambda^2}{8 \cdot \pi} \tag{5.80}$$

According to (5.68), the gain of the Hertzian dipole, regardless of whether the Hertzian dipole is used as a transmitting or a receiving antenna, is

$$G_{\text{Hertz}} = \frac{3}{2}$$

The reference antenna for the gain is the isotropic radiator, i.e.,

$$G_{\text{Hertz}} = \frac{A_{\text{eff/Hertz}}}{A_{\text{eff/isotrop}}}$$

The effective area of an isotropic transmitting antenna, or of an isotropic receiving antenna, is

$$A_{\text{eff/iso}} = \frac{A_{\text{eff/Hertz}}}{G_{\text{Hertz}}} = \frac{\lambda^2}{4 \cdot \pi} \tag{5.81}$$

Generally, the gain G_{ant} of an antenna is

$$G_{\text{ant}} = \frac{A_{\text{eff/ant}}}{A_{\text{eff/iso}}} = A_{\text{eff/ant}} \cdot \frac{4 \cdot \pi}{\lambda^2} \tag{5.82}$$

In the case of the Hertzian dipole, the effective area is not related to a real surface; it is purely a theoretical physical quantity. In the case of aperture antennas, e.g., a parabolic dish antenna shown in Fig. 5.12, the effective area can be related to the geometric dimensions. In the case of a parabolic antenna, the shadow surface is a circular surface and is called aperture.

The aperture of such an antenna with the diameter D is

$$A_{\text{ap}} = \frac{\pi \cdot D^2}{4} \tag{5.83}$$

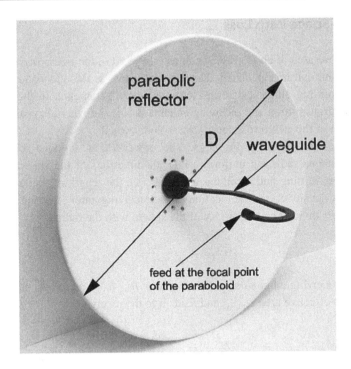

Fig. 5.12 Paraboloid dish antenna

To determine the gain of the antenna on the basis of the aperture, the aperture is weighted with the so-called efficiency q. The antenna efficiency q takes into account that the feed does not generate a constant power density across the aperture. This is desirable in order to achieve a radiation pattern with high attenuation of the side lobes. The value of the antenna efficiency ranges between 0.5 and 0.7, depending on the antenna design. Consequently, the gain of an aperture antenna is

$$G_{\text{ant}} = \frac{A_{\text{eff/ant}}}{A_{\text{eff/iso}}} = \frac{\frac{\pi \cdot D^2}{4} \cdot q}{\frac{\lambda^2}{4 \cdot \pi}}$$

$$G_{\text{ant}} = \frac{\pi^2 \cdot D^2 \cdot q}{\lambda^2} \tag{5.84}$$

or

$$g_{\text{ant}} = 10 \cdot \log_{10} \frac{\pi^2 \cdot D^2 \cdot q}{\lambda^2} \tag{5.85}$$

From (5.84), it also becomes evident why the gain of an antenna is inversely proportional to the square of the wavelength λ. At constant wavelength, the gain of an antenna increases with a larger aperture surface.

5.2   Free-Space Path Loss

With the effective area of an isotropic antenna (see (5.81)), the free-space loss of a radio transmission path can be calculated. The free-space loss is the attenuation between the isotropic transmitting antenna and an isotropic receiving antenna in the case that the space between transmitting and receiving antenna is free, i.e., there is visual contact and the propagation of the electromagnetic wave is undisturbed.

This situation is shown in Fig. 5.13. The power P_{tr} is supplied to the isotropic transmitting antenna radiating uniformly into all directions of the space. The distance between the transmitting and receiving antennas, i.e., the length of the radio path, is d. The power flux density S_{rec} at the location of the receiving antenna is equal to the transmitting power P_{tr} divided by the surface of the sphere with the radius d

$$S_{rec} = \frac{P_{tr}}{4 \cdot \pi \cdot d^2} \tag{5.86}$$

The isotropic receiving antenna absorbs the power P_{rec} from the power flux density S_{rec} according to its effective area $A_{eff/iso}$ and feeds it to the receiver

$$P_{rec} = S_{rec} \cdot A_{eff/iso} = S_{rec} \cdot \frac{\lambda^2}{4 \cdot \pi} = \frac{P_{tr}}{4 \cdot \pi \cdot d^2} \cdot \frac{\lambda^2}{4 \cdot \pi} = P_{tr} \cdot \left(\frac{\lambda}{4 \cdot \pi \cdot d} \right)^2$$

The attenuation between the isotropic transmitting antenna and the isotropic receiving antenna is called the free-space loss a_0.

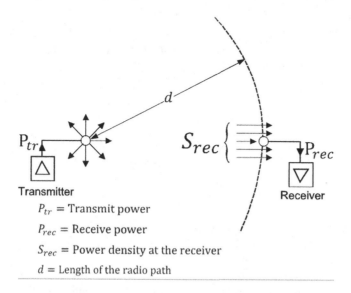

P_{tr} = Transmit power

P_{rec} = Receive power

S_{rec} = Power density at the receiver

d = Length of the radio path

Fig. 5.13 Radio path with isotropic antennas as transmit and receive antennas

$$a_0 = 10 \cdot \log_{10} \frac{P_{\text{tr}}}{P_{\text{rec}}} \, \text{dB} = 20 \cdot \log_{10} \frac{4 \cdot \pi \cdot d}{\lambda} \, \text{dB}$$

or

$$a_0 = \left[20 \cdot \log_{10} (4 \cdot \pi) - 20 \cdot \log_{10} \lambda + 20 \cdot \log_{10} d \right] \text{dB}$$

With $f = \frac{c_0}{\lambda}$

$$a_0 = \left(20 \cdot \log_{10} \frac{4 \cdot \pi}{c_0} + 20 \cdot \log_{10} f + 20 \cdot \log_{10} d \right) \text{dB}$$

For a frequency f in GHz and d in km with $c_0 = 2{,}998 \cdot 10^8$ m/s, the free-space loss is

$$a_0 = \left(92.4 + 20 \cdot \log_{10} \frac{f}{\text{GHz}} + 20 \cdot \log_{10} \frac{d}{\text{km}} \right) \text{dB} \tag{5.87}$$

The condition of unobstructed propagation is generally fulfilled for microwave links. Parabolic antennas with gain values between 30 and 45 dB are used for this type of radio link. In contrast to the free-space loss, the radio field loss includes the antenna gain of the transmitting and receiving antennas. The radio field loss is consequently lower than the free-space loss by the sum of the antenna gains of the transmitting and receiving antennas. With the gain g_{tr} of the transmitting antenna

$$g_{\text{tr}} = 10 \cdot \log_{10} G_{\text{tr}}$$

and the gain g_{rec} of the receiving antenna

$$g_{\text{rec}} = 10 \cdot \log_{10} G_{\text{rec}}$$

the radio field loss a_{RF} is

$$a_{\text{RF}} = a_0 - g_{\text{tr}} - g_{\text{rec}} \tag{5.88}$$

Appendix: Verification of the Calculation Rules for Vector Analysis

6.1 Calculation Rules

The following calculation rules for vector analysis were used in this volume (see (3.86) to (3.89)):

φ

$$\nabla \times \nabla\varphi = 0 \tag{3.86}$$

Div

$$\nabla \cdot (\nabla \times \vec{V}) = 0 \tag{3.87}$$

$$\nabla \times \left(\nabla \times \vec{V}\right) = \nabla\left(\nabla \cdot \vec{V}\right) - \nabla^2 \vec{V} \tag{3.88}$$

$$\nabla \cdot \left(\vec{V} \times \vec{B}\right) = \vec{B} \cdot \left(\nabla \times \vec{V}\right) - \vec{V} \cdot \left(\nabla \times \vec{B}\right) \tag{3.89}$$

6.2 Proofs

The equations will be verified for Cartesian coordinates.

6.2.1 Proof of Eq. (3.86)

φ

$$\nabla \times \nabla\varphi = 0$$

From (1.7) in conjunction with (3.75), we obtain

$$\nabla \times \nabla\varphi = \begin{vmatrix} \vec{e}_x \vec{e}_y & \vec{e}_z \\ \dfrac{\partial}{\partial x} \dfrac{\partial}{\partial y} & \dfrac{\partial}{\partial z} \\ \dfrac{\partial\varphi}{\partial x} \dfrac{\partial\varphi}{\partial y} & \dfrac{\partial\varphi}{\partial z} \end{vmatrix}$$

$$\nabla \times \nabla\varphi = \vec{e}_x\left(\frac{\partial^2\varphi}{\partial y\partial z} - \frac{\partial^2\varphi}{\partial z\partial y}\right) + \vec{e}_y\left(\frac{\partial^2\varphi}{\partial z\partial x} - \frac{\partial^2\varphi}{\partial x\partial z}\right) + \vec{e}_z\left(\frac{\partial^2\varphi}{\partial x\partial y} - \frac{\partial^2\varphi}{\partial y\partial x}\right)$$

According to Schwarz's theorem, the order of partial derivatives of functions is interchangeable if the functions and their partial derivatives are continuous. Therefore, according to (3.86)

$$\nabla \times \nabla\varphi = 0$$

6.2.2   Proof of Eq. (3.87)

Div

$$\nabla \cdot (\nabla \times \vec{V}) = 0 \tag{3.87}$$

After (2.28) and (2.29) with (3.76)

$$\nabla \cdot \left(\nabla \times \vec{V}\right) = \frac{\partial}{\partial x}\left(\frac{\partial V_z}{\partial y} - \frac{\partial V_y}{\partial z}\right) + \frac{\partial}{\partial y}\left(\frac{\partial V_x}{\partial z} - \frac{\partial V_z}{\partial x}\right) + \frac{\partial}{\partial z}\left(\frac{\partial V_y}{\partial x} - \frac{\partial V_x}{\partial y}\right)$$

$$\nabla \cdot \left(\nabla \times \vec{V}\right) = \frac{\partial^2 V_z}{\partial x\partial y} - \frac{\partial^2 V_y}{\partial x\partial z} + \frac{\partial^2 V_x}{\partial y\partial z} - \frac{\partial^2 V_z}{\partial y\partial x} + \frac{\partial^2 V_y}{\partial z\partial x} - \frac{\partial^2 V_x}{\partial z\partial y} = 0$$

6.2.3   Proof of Eq. (3.88)

$$\nabla \times \left(\nabla \times \vec{V}\right) = \nabla\left(\nabla \cdot \vec{V}\right) - \nabla^2 \vec{V}$$

According to (3.73)

$$\nabla \times \vec{V} = \begin{vmatrix} \vec{e}_x & \vec{e}_y & \vec{e}_z \\ \dfrac{\partial}{\partial x} & \dfrac{\partial}{\partial y} & \dfrac{\partial}{\partial z} \\ V_x & V_y & V_z \end{vmatrix} = \vec{e}_x \cdot \left(\frac{\partial V_z}{\partial y} - \frac{\partial V_y}{\partial z}\right) + \vec{e}_y \cdot \left(\frac{\partial V_x}{\partial z} - \frac{\partial V_z}{\partial x}\right) + \vec{e}_z \cdot \left(\frac{\partial V_y}{\partial x} - \frac{\partial V_x}{\partial y}\right)$$

$$\nabla \times \left(\nabla \times \vec{V}\right) = \nabla \times \begin{vmatrix} \vec{e}_x & \vec{e}_y & \vec{e}_z \\ \dfrac{\partial}{\partial x} & \dfrac{\partial}{\partial y} & \dfrac{\partial}{\partial z} \\ V_x & V_y & V_z \end{vmatrix} = \begin{vmatrix} \vec{e}_x & \vec{e}_y & \vec{e}_z \\ \dfrac{\partial}{\partial x} & \dfrac{\partial}{\partial y} & \dfrac{\partial}{\partial z} \\ \left(\dfrac{\partial V_z}{\partial y} - \dfrac{\partial V_y}{\partial z}\right) & \left(\dfrac{\partial V_x}{\partial z} - \dfrac{\partial V_z}{\partial x}\right) & \left(\dfrac{\partial V_y}{\partial x} - \dfrac{\partial V_x}{\partial y}\right) \end{vmatrix}$$

$$\nabla \times \left(\nabla \times \vec{V}\right) = \vec{e}_x \cdot \left[\frac{\partial}{\partial y}\left(\frac{\partial V_y}{\partial x} - \frac{\partial V_x}{\partial y}\right) - \frac{\partial}{\partial z}\left(\frac{\partial V_x}{\partial z} - \frac{\partial V_z}{\partial x}\right)\right]$$
$$+ \vec{e}_y \cdot \left[\frac{\partial}{\partial z}\left(\frac{\partial V_z}{\partial y} - \frac{\partial V_y}{\partial z}\right) - \frac{\partial}{\partial x}\left(\frac{\partial V_y}{\partial x} - \frac{\partial V_x}{\partial y}\right)\right]$$
$$+ \vec{e}_z \cdot \left[\frac{\partial}{\partial x}\left(\frac{\partial V_x}{\partial z} - \frac{\partial V_z}{\partial x}\right) - \frac{\partial}{\partial y}\left(\frac{\partial V_z}{\partial y} - \frac{\partial V_y}{\partial z}\right)\right]$$

$$\nabla \times \left(\nabla \times \vec{V}\right) = \vec{e}_x \cdot \left(\frac{\partial^2 V_y}{\partial x \cdot \partial y} - \frac{\partial^2 V_x}{\partial y^2} - \frac{\partial^2 V_x}{\partial z^2} + \frac{\partial^2 V_z}{\partial x \cdot \partial z}\right)$$
$$+ \vec{e}_y \cdot \left(\frac{\partial^2 V_z}{\partial y \cdot \partial z} - \frac{\partial^2 V_y}{\partial z^2} - \frac{\partial^2 V_y}{\partial x^2} + \frac{\partial^2 V_x}{\partial x \cdot \partial y}\right) \qquad (6.1)$$
$$+ \vec{e}_z \cdot \left(\frac{\partial^2 V_x}{\partial x \cdot \partial z} - \frac{\partial^2 V_z}{\partial x^2} - \frac{\partial^2 V_z}{\partial y^2} + \frac{\partial^2 V_y}{\partial y \cdot \partial z}\right)$$

According to (1.7):

$$\nabla\left(\nabla \cdot \vec{V}\right) = \frac{\partial}{\partial x}\left(\nabla \cdot \vec{V}\right) \cdot \vec{e}_x + \frac{\partial}{\partial y}\left(\nabla \cdot \vec{V}\right) \cdot \vec{e}_y + \frac{\partial}{\partial z}\left(\nabla \cdot \vec{V}\right) \cdot \vec{e}_z$$

With (2.28)

$$\nabla\left(\nabla \cdot \vec{V}\right) = \frac{\partial}{\partial x}\left[\left(\frac{\partial V_x}{\partial x} + \frac{\partial V_y}{\partial y} + \frac{\partial V_z}{\partial z}\right) \cdot \vec{e}_x\right]$$
$$+ \frac{\partial}{\partial y}\left[\left(\frac{\partial V_x}{\partial x} + \frac{\partial V_y}{\partial y} + \frac{\partial V_z}{\partial z}\right) \cdot \vec{e}_y\right] + \frac{\partial}{\partial z}\left[\left(\frac{\partial V_x}{\partial x} + \frac{\partial V_y}{\partial y} + \frac{\partial V_z}{\partial z}\right) \cdot \vec{e}_z\right]$$

and

$$\nabla\left(\nabla \cdot \vec{V}\right) = \left(\frac{\partial^2 V_x}{\partial x^2} + \frac{\partial^2 V_y}{\partial x \cdot \partial y} + \frac{\partial^2 V_z}{\partial x \cdot \partial z}\right) \cdot \vec{e}_x + \left(\frac{\partial^2 V_x}{\partial x \cdot \partial y} + \frac{\partial^2 V_y}{\partial y^2} + \frac{\partial^2 V_z}{\partial y \cdot \partial z}\right) \cdot \vec{e}_y$$
$$+ \left(\frac{\partial^2 V_x}{\partial x \cdot \partial z} + \frac{\partial^2 V_y}{\partial y \cdot \partial z} + \frac{\partial^2 V_z}{\partial z^2}\right) \cdot \vec{e}_z$$

$$(6.2)$$

The sum of the gray marked expressions in (6.1) and (6.2) equals zero. Thus

$$\nabla\left(\nabla \cdot \vec{V}\right) - \nabla \times \left(\nabla \times \vec{V}\right) = \left(\frac{\partial^2 V_x}{\partial x^2} + \frac{\partial^2 V_x}{\partial y^2} + \frac{\partial^2 V_x}{\partial z^2}\right) \cdot \vec{e}_x$$
$$+ \left(\frac{\partial^2 V_y}{\partial y^2} + \frac{\partial^2 V_y}{\partial z^2} + \frac{\partial^2 V_y}{\partial x^2}\right) \cdot \vec{e}_y + \left(\frac{\partial^2 V_z}{\partial z^2} + \frac{\partial^2 V_z}{\partial x^2} + \frac{\partial^2 V_z}{\partial y^2}\right) \cdot \vec{e}_z$$

$$(6.3)$$

After (4.72)

$$\nabla^2 \vec{V} = \left(\nabla^2 V_x\right) \cdot \vec{e}_x + \left(\nabla^2 V_y\right) \cdot \vec{e}_y + \left(\nabla^2 V_z\right) \cdot \vec{e}_z$$

In this equation, the operator ∇^2 is defined as (see also (2.43))

$$\nabla^2 = \frac{\partial^2}{\partial x^2} + \frac{\partial^2}{\partial y^2} + \frac{\partial^2}{\partial z^2}$$

From both equations, we get

$$\nabla^2 \vec{V} = \left(\frac{\partial^2 V_x}{\partial x^2} + \frac{\partial^2 V_y}{\partial y^2} + \frac{\partial^2 V_z}{\partial z^2}\right) \cdot \vec{e}_x + \left(\frac{\partial^2 V_x}{\partial x^2} + \frac{\partial^2 V_y}{\partial y^2} + \frac{\partial^2 V_z}{\partial z^2}\right) \cdot \vec{e}_y$$
$$+ \left(\frac{\partial^2 V_x}{\partial x^2} + \frac{\partial^2 V_y}{\partial y^2} + \frac{\partial^2 V_z}{\partial z^2}\right) \cdot \vec{e}_z$$

(6.4)

A comparison of (6.3) and (6.4) gives the following relationship:

$$\nabla^2 \vec{V} = \nabla \left(\nabla \cdot \vec{V}\right) - \nabla \times \left(\nabla \times \vec{V}\right)$$

This proves the validity of (3.88).

6.2.4 Proof of Eq. (3.89)

$$\nabla \cdot \left(\vec{V} \times \vec{B}\right) = \vec{B} \cdot \left(\nabla \times \vec{V}\right) - \vec{V} \cdot \left(\nabla \times \vec{B}\right)$$

The cross product of two vectors is

$$\vec{V} \times \vec{B} = \begin{vmatrix} \vec{e}_x & \vec{e}_y & \vec{e}_z \\ V_x & V_y & V_z \\ B_x & B_y & B_z \end{vmatrix}$$

$$\vec{V} \times \vec{B} = \vec{e}_x \cdot (V_y \cdot B_z - V_z \cdot B_y) + \vec{e}_y \cdot (V_z \cdot B_x - V_x \cdot B_z) + \vec{e}_z \cdot (V_x \cdot B_y - V_y \cdot B_x)$$

$$\nabla \cdot \left(\vec{V} \times \vec{B}\right) = \frac{\partial}{\partial x}(V_y \cdot B_z - V_z \cdot B_y) + \frac{\partial}{\partial y}(V_z \cdot B_x - V_x \cdot B_z) + \frac{\partial}{\partial z}(V_x \cdot B_y - V_y \cdot B_x)$$

With the product rule

$$\nabla \cdot \left(\vec{V} \times \vec{B}\right) = \frac{\partial V_y}{\partial x} \cdot B_z + \frac{\partial B_z}{\partial x} \cdot V_y - \frac{\partial V_z}{\partial x} \cdot B_y - \frac{\partial B_y}{\partial x} \cdot V_z$$
$$+ \frac{\partial V_z}{\partial y} \cdot B_x + \frac{\partial B_x}{\partial y} \cdot V_z - \frac{\partial V_x}{\partial y} \cdot B_z - \frac{\partial B_z}{\partial y} \cdot V_x$$
$$+ \frac{\partial V_x}{\partial z} \cdot B_y + \frac{\partial B_y}{\partial z} \cdot V_x - \frac{\partial V_y}{\partial z} \cdot B_x - \frac{\partial B_x}{\partial z} \cdot V_y$$

(6.5)

also

$$\nabla \times \overrightarrow{V} = \begin{vmatrix} \overrightarrow{e}_x & \overrightarrow{e}_y & \overrightarrow{e}_z \\ \frac{\partial}{\partial x} & \frac{\partial}{\partial y} & \frac{\partial}{\partial z} \\ V_x & V_y & V_z \end{vmatrix} = \overrightarrow{e}_x \cdot \left(\frac{\partial V_z}{\partial y} - \frac{\partial V_y}{\partial z} \right) + \overrightarrow{e}_y \cdot \left(\frac{\partial V_x}{\partial z} - \frac{\partial V_z}{\partial x} \right) + \overrightarrow{e}_z \cdot \left(\frac{\partial V_y}{\partial x} - \frac{\partial V_x}{\partial y} \right)$$

and

$$\overrightarrow{B} \cdot \left(\nabla \times \overrightarrow{V} \right) = B_x \cdot \left(\frac{\partial V_z}{\partial y} - \frac{\partial V_y}{\partial z} \right) + B_y \cdot \left(\frac{\partial V_x}{\partial z} - \frac{\partial V_z}{\partial x} \right) + B_z \cdot \left(\frac{\partial V_y}{\partial x} - \frac{\partial V_x}{\partial y} \right)$$

$$\overrightarrow{B} \cdot \left(\nabla \times \overrightarrow{V} \right) = \frac{\partial V_z}{\partial y} \cdot B_x - \frac{\partial V_y}{\partial z} \cdot B_x + \frac{\partial V_x}{\partial z} \cdot B_y - \frac{\partial V_z}{\partial x} \cdot B_y + \frac{\partial V_y}{\partial x} \cdot B_z - \frac{\partial V_x}{\partial y} \cdot B_z$$

(6.6)

In analogy to this equation, we have

$$-\overrightarrow{V} \cdot \left(\nabla \times \overrightarrow{B} \right) = -\frac{\partial B_z}{\partial y} \cdot V_x + \frac{\partial B_y}{\partial z} \cdot V_x - \frac{\partial B_x}{\partial z} \cdot V_y + \frac{\partial B_z}{\partial x} \cdot V_y - \frac{\partial B_y}{\partial x} \cdot V_z + \frac{\partial B_x}{\partial y} \cdot V_z$$

(6.7)

In (6.5), (6.6), and (6.7), the derivatives $\partial/\partial x$ are marked gray. We can see that the marked relations in (6.5) correspond to the marked relations in (6.6) and (6.7). The same applies to the derivatives $\partial/\partial y$ and $\partial/\partial z$. This confirms the validity of (3.89).

References

1. Bronstein, I.N., Semendjajew, K.A.: Taschenbuch der Mathematik, 2nd edn. Verlag Harri Deutsch, Frankfurt a. M. (1962)
2. Kark, K.: Antennen und Strahlungsfelder, 2nd edn. Friedrich Vieweg & Sohn Verlag, Wiesbaden (2006)
3. Kröger, R., Unbehauen, R.: Elektrodynamik. Teubner, Stuttgart, (1990)
4. Küpfmüller, K., Kohn, G.: Theoretische Elektrotechnik und Elektronik, 14th improved edn. Springer, Berlin (1993)
5. Lehner, G.: Electromagnetic field theory. Springer, Berlin (2006)
6. Simonyi, K.: Theoretische Elektrotechnik. VEB Deutscher Verlag der Wissenschaften, Berlin (1956)
7. Stöcker, H.: Taschenbuch mathematischer Formeln und moderner Verfahren, 2. überarbeitete Aufl. Verlag Harry Deutsch, Frankfurt a. M. (1992)
8. Zinke, O., Brunswig, H.: Lehrbuch der Hochfrequenztechnik. Springer, Berlin (1965)

Animations of the wave propagation of the Hertzian dipole:

http://www-tet.ee.tu-berlin.de/Animationen/HertzscherDipol1/
http://www.mikomma.de/fh/eldy/hertz.html
https://www.leifiphysik.de/elektrizitaetslehre/elektromagnetische-wellen/versuche/dipolstrahlung-animation
https://www.didaktik.physik.uni-muenchen.de/multimedia/programme_applets/e_lehre/dipolstrahlung/

Further Literature

Guru, B., Hiziroğlu. H.: Electromagnetic Field Theory Fundamentals, Cambridge University Press, ISBN 978-0-521-83016-4 (2009)
Ramo, Simon, Whinnery, John R., and Van Duzer, Theodore: Fields and Waves in Communication Electronics, John Wiley & Sons Inc., ISBN 978-0-471-58551-0
Henke, H.: Elektromagnetische Felder – Theorie und Anwendung, 4. bearbeitete Aufl. Springer-Verlag, ISBN 978-3-462-19745-1 (2011)
Kark, K.: Antennen und Strahlungsfelder, 5. Aufl. Springer-Verlag, ISBN 978-3-658-03615-7 (2013)

© Springer Fachmedien Wiesbaden GmbH, part of Springer Nature 2020
J. Donnevert, *Maxwell´s Equations,* https://doi.org/10.1007/978-3-658-29376-5

Klingbeil, H: Grundlagen der elektromagnetischen Feldtheorie, 3. Aufl. Springer Spektrum, ISBN
    978-3-662-56599-5
Lehner, G.: Elektromagnetische Feldtheorie für Ingenieure und Physiker, 7. bearbeitete Auflage,
    Springer-Verlag, ISBN 978-3-642-13041-0 (2010)
Fließbach, T.: Elektrodynamik, Lehrbuch zur Theoretischen Physik II. Verlag Springer Spektrum,
    ISBN 978-3-8274-3036-6

Index

© Springer Fachmedien Wiesbaden GmbH, part of Springer Nature 2020
J. Donnevert, *Maxwell´s Equations,* https://doi.org/10.1007/978-3-658-29376-5

Printed in the United States
By Bookmasters